A Text Book Of

PETROLEUM AND PETROCHEMICAL TECHNOLOGY

(22611)

Semester - VI

THIRD YEAR DIPLOMA COURSE IN CHEMICAL ENGINEERING GROUP

As Per MSBTE's 'I' Scheme Syllabus

YOGESH A. KARPE
B.E. (Chem.), M.Tech. (Chem. Tech.)
Assistant Professor,
Department of Chemical Engineering,
M.G.M. College of Engineering and Technology,
Navi Mumbai.

NIRALI PRAKASHAN
ADVANCEMENT OF KNOWLEDGE

N4408

Petroleum and Petrochemical Technology (Sem. - VI) **ISBN 978-93-89825-88-6**

First Edition : January 2020

Published By :
NIRALI PRAKASHAN
Abhyudaya Pragati, 1312, Shivaji Nagar
Off J.M. Road, PUNE – 411005
Tel - (020) 25512336/37/39, Fax - (020) 25511379
Email : niralipune@pragationline.com

➢ DISTRIBUTION CENTRES

PUNE

Nirali Prakashan (For orders within Pune) : 119, Budhwar Peth, Jogeshwari Mandir Lane, Pune 411002, Maharashtra
Tel : (020) 2445 2044, Fax : (020) 2445 1538; Mobile : 9657703145
Email : bookorder@pragationline.com, niralilocal@pragationline.com

Nirali Prakashan (For orders outside Pune) : S. No. 28/27, Dhyari, Near Pari Company, Pune 411041
Tel : (020) 24690204 Fax : (020) 24690316; Mobile : 9657703143
Email : dhyari@pragationline.com, bookorder@pragationline.com

MUMBAI

Nirali Prakashan : 385, S.V.P. Road, Rasdhara Co-op. Hsg. Society Ltd.,
Girgaum, Mumbai 400004, Maharashtra; Mobile : 9320129587
Tel : (022) 2385 6339 / 2386 9976, Fax : (022) 2386 9976
Email : niralimumbai@pragationline.com

➢ DISTRIBUTION BRANCHES

JALGAON

Nirali Prakashan : 34, V. V. Golani Market, Navi Peth, Jalgaon 425001,
Maharashtra, Tel : (0257) 222 0395, Mob : 94234 91860
Email : niralijalgoan@pragationline.com

KOLHAPUR

Nirali Prakashan : New Mahadvar Road, Kedar Plaza, 1st Floor Opp. IDBI Bank
Kolhapur 416 012, Maharashtra. Mob : 9850046155
Email : niralikolhapur@pragationline.com

NAGPUR

Nirali Prakashan : Above Maratha Mandir, Shop No. 3, First Floor,
Rani Jhanshi Square, Sitabuldi, Nagpur 440012, Maharashtra
Tel : (0712) 254 7129; Email : niralinagpur@pragationline.com

DELHI

Nirali Prakashan : 4593/15, Basement, Agarwal Lane, Ansari Road, Daryaganj
Near Times of India Building, New Delhi 110002 Mob : 08505972553
Email : niralidelhi@pragationline.com

BANGALURU

Nirali Prakashan : Maitri Ground Floor, Jaya Apartments, No. 99, 6th Cross, 6th Main,
Malleswaram, Bangaluru 560 003, Karnataka
Mob : +91 9449043034
Email: niralibangalore@pragationline.com

Dedicated To,

"My Beloved Wife and Charming Son"

Preface ...

The importance of **Petroleum and Petrochemical Technology** is well-known in chemical engineering field. No availability of any material or any book of **Petroleum and Petrochemical Technology**, and students demand for study materials of the subject inspired me to write this book. The book is structured to cover the key aspects of the subject **Petroleum and Petrochemical Technology** as per new revised syllabus of MSBTE (I-scheme).

The book has simple, lucid language to explain fundamentals of this subject. The book provides logical methods of explaining various concepts and illustrations of important topics. Each chapter is well supported with necessary illustrations, examples and well defined labelled diagrams. All chapters in this book are arranged in a proper sequence that permits each topic to build upon earlier studies. All care has been taken to make students easy and comfortable in understanding the basic concepts of this subject.

The book not only covers the entire scope of the subject but explains the basics of the subject. This makes the understanding of this subject clearer and makes it more interesting. The book will be very useful not only to the students but also to subject teachers. The students have to omit nothing and possibly have to cover nothing more.

I wish to express my heartly and profound thanks to all those who helped in making this book a reality. Much needed moral support and encouragement is provided on numerous occasions by my all friends and my family.

I wish to thanks the Publisher **Mr. Dineshbhai Furia, Mr. Jignesh Furia** and the staff of Nirali Prakashan, specially Mr. Ilyas Shaikh, Mrs. Manasi Pingle, Ms. Chaitali Takale and Mr. Ravindra Walodare who have immense pain to get this book in time with quality printing.

Any suggestion for the improvement of the book will be acknowledged and well appreciated.

Yogesh A. Karpe

✍ ✍ ✍

Syllabus ...

1. Petroleum and Refinery (Marks 08 - Hours 06)

1.1 Petroleum: Occurrence and History, Crude oil reserves in India.

1.2 Composition of petroleum and petroleum products.

1.3 Classification of petroleum.

1.4 Indian petroleum refinery and the different sections and equipment.

1.5 Refinery products.

2. Distillation of Crude Oil (Marks 18 - Hours 12)

2.1 Primary treatment of crude: Dehydration and desalting of crude.

2.2 Atmospheric distillation of crude.

2.3 Vacuum distillation of crude.

2.4 Fractions obtained from crude oil with their boiling range and uses.

2.5 Properties of petroleum products: Octane number, cetane number, flash point, fire point, aniline point, pour point, cloud point, drop point, calorific value, carbon residue.

2.6 Pensky Marten's apparatus, Abel's apparatus, cloud and pour point apparatus, drop point apparatus, smoke point apparatus, aniline point apparatus, Conradson apparatus.

2.7 Measurement of properties of petroleum products: Gasoline (ASTM distillation set-up, specific gravity), Kerosene (flash and fire point, smoke point, aniline point) Diesel (cloud and pour point, aniline point, flash point, viscosity) wax (drop point) Lubricating oil.

3. Processes in Refinery (Marks 18 - Hours 12)

3.1 Cracking: Definition, Types of cracking (thermal and catalytic).

3.2 Thermal cracking: Visbreaking and delayed coking. (Flow sheet and reaction of following).

3.3 Catalytic cracking: Fixed and fluidized bed catalytic cracking.

3.4 Reforming: Catalytic reforming (platforming).

3.5 Hydrogenation, Hydrocracking and Isomerization.

3.6 Alkylation: Sulfuric acid and (Hydrofloric acid) HF alkylation process.

3.7 Esterification and Hydration.

3.8 Waste treatment: Oil removal, Organic content removal, Solid and hazardous waste treatment.

4. Petrochemicals (From C_1 to C_3) (Marks 16 - Hours 10)

4.1 Petrochemical industries in India, List of chemicals from C_1, Manufacturing of Methanol and Formaldehyde.

4.2 Esterification process.

4.3 List of chemicals from C_2, Manufacturing of Ethanol, Vinyl chloride, Ethylene oxide and Styrene from C_2.

4.4 List of chemicals from C_3, Manufacturing of Propylene Oxide and Acetaldehyde.

5. Petrochemicals From C_4 and Aromatics (Marks 10 - Hours 08)

5.1 Chemicals from C_4 and Manufacturing of Butadiene and Methyl Tertiary Butyl Ether

5.2 List of chemicals from aromatics

5.3 Production and separation of Benzene Toluene and Xylene

5.4 Manufacture of aniline from phenol

5.5 Manufacture of benzoic acid from toluene

Contents ...

Chapter **1 ...**

PETROLEUM AND REFINERY

Weightage of Marks = 08, Teaching Hours = 06

Syllabus

1.1 Petroleum: Occurrence and History, Crude oil reserves in India.

1.2 Composition of petroleum and petroleum products.

1.3 Classification of petroleum.

1.4 Indian petroleum refinery and the different sections and equipment.

1.5 Refinery products.

About this Chapter

At the end of this chapter, students will be able to:

- State the composition of the specified petroleum product.
- Classify petroleum on the basis of given criteria.
- Describe with sketches the function of the specified petroleum refinery equipment/component.
- State the properties of the specified refinery products.

1.1 PETROLEUM: OCCURRENCE AND HISTORY [S-16, 18; W-17]

Petroleum has an organic origin and is found in sedimentary basins, shallow depressions and in the seas (past and present). Most of the oil reserves in India are associated with anticlines and fault traps in the sedimentary rock formations of tertiary times, about 3 million years ago. Some recent sediment, less than one million years also show evidence of incipient oil.

Oil and natural gas originated from animal or vegetable matter contained in shallow marine sediments, such as sands, silts and clays deposited during the periods when land and aquatic life was abundant in various forms, especially the minor microscopic forms of flora and fauna.

Conditions for oil formation were favourable especially in the lower and middle Tertiary period. Dense forests and sea organisms flourished in the gulfs, estuaries, deltas and the land surrounding them during this period. The decomposition of organic matter in the sedimentary rocks has led to the formation of oil. Though oil is mainly found in sedimentary rocks, all sedimentary rocks do not contain oil.

An oil reservoir must have three pre-requisite conditions: (i) porosity so as to accommodate sufficiently large amounts of oil; (ii) permeability to discharge oil and/or gas when well has been drilled; (iii) the porous sand beds sandstone, conglomerates of fissured limestone containing oil should be capped by impervious beds so that oil does not dissipate by percolation in the surrounding rocks.

Oil on a commercial scale is usually found where the sedimentary rock strata are inclined and folded; in a sort of chamber or reservoir, in the highest possible situation e.g. crests of anticlines. Normally, oil is associated with water. Being lighter than water (specific gravity of 0.8 to 0.98), it collects in the anticlines or fault traps above the surface of water. Gas is still lighter and occurs above oil. Thus on drilling an oil well, one finds gas followed by oil, although gas seepage is not always a sure indication of an oil reservoir.

As already mentioned, oil as well as natural gas in India occur in sedimentary rocks. About 14.1 lakh sq. km. or about 42 per cent of the total area of the country is covered with sedimentary rocks out of which about 10 lakh sq. km. form marine basins of Mesozoic and Tertiary times.

Besides, the country has offshore areas having Mesozoic and Tertiary rocks of marine origin covering an area of 2.5 lakh sq. km. upto a depth of 100 metre and another area of 0.7 lakh sq. km. upto a depth between 100 and 200 metre. Thus, the total continental shelf of probable oil bearing rocks amounts to 3.2 lakh sq. km.

The total sedimentary area including both on shore and offshore comprises 27 basins. The geological and geophysical studies have been conducted in 14 basins while exploratory drilling has been done in 9 basins. Mumbai High, the Khambhat Gulf and the Assam are the most productive areas.

History :

The history of crude oil dates back to the 3rd or 4th century A.D when the presence of oil was first discovered in China. The oil that the early Chinese people found was to have extremely good medicinal value and was used in the salt form. To extract that oil from under the earth's crust, first oil wells of around 243 metres were dug up in that region with the help of bamboo poles possessing metal tools at their end. The crude oil was also used for the lighting purposes in Ancient Persia.

When the city of Baghdad was constructed in the 8th century, the streets of that city were paved with tar that was easily available in the natural oil fields in that region. With the time, man discovered new and diversified uses of oil and hence the discovery of new oil fields became an important requirement. People got to know the vast amount of oil that was hidden under the earth's surface as many geographers started predicting it in that time.

All this time oil had a limited use until in 19th century; the process of distillation of kerosene with the help of coal and rock oil was invented. Production of oil became commercialized and it started an era of establishment of oil refineries throughout the world. One of the first refineries was established at Baku in Russia and it became the world's largest oil producing refinery. A Russian engineer F.N. Semyonov built the first ever modern oil well at the same place in 1848. Uptill 1950s coal had dominance among the primary energy constituents but crude oil took over the leadership in a short span of time and has still maintained its reputation.

History of Crude Oil in India :

India was not known to the world in the context of crude oil and it's by-product production. As late as in 1889, the presence of oil in India was discovered in Digboi in Assam. First crude oil refinery in India was set up in Digboi in 1901. Then the exploration and production activities were limited to the North Eastern part of the country. In 1958 and 1974, two more places for crude oil production were identified namely Cambay onshore basin and Bombay offshore basin. Initially, the major international companies were given the job to explore and produce oil in the country but after the shock in oil prices in 1973, whole of the sector was nationalized.

The first oil deposits in India were discovered in 1889 near the town of Digboi in the state of Assam.

The natural gas industry in India began in the 1960s with the discovery of gas fields in Assam and Gujarat. Natural gas gained further significance after the discovery of large reserves in the South Basin fields by ONGC in the 1970s.

1.1.1 Crude Oil Reserves in India

Reserves:

Although India has vast areas covered by sedimentary rocks, structures containing oil are not in proportion to the expanses of these rocks and are found in limited situations. The Indian Mineral Year Book 1982 estimated a reserve of 468 million tonnes of which 328 million tonnes was available in Mumbai High. In 1984, the reserves were estimated at 500 million tonnes.

The Indian Petroleum and Natural Gas Statistics put the total reserves of crude oil at 581.43 million tonnes in 1986-87. The prognosticated hydrocarbon resource base in Indian sedimentary basins including deep water has been estimated at about 28 billion tonnes.

Of this only about one-fourth i.e., 7.2 billion tonnes of in place hydrocarbon reserves have been established as on 1st April, 2002. About 70 per cent of the established hydrocarbon reserves is oil and rest is gas. The recoverable hydrocarbon reserves are of the order of 2.6 billion tonnes.

As on 31st March 2018, India had estimated crude oil reserves of 594.49 million tonnes (MT), declining by 1.59% from the previous year. The largest reserves are found in the Western Offshore (40%), and Assam (27%). The estimated reserves of natural gas in India as on 31st March 2018 was 1,339.57 billion cubic metres (BCM), increasing by 3.87% from the previous year. The largest reserves of natural gas are located in the Eastern Offshore (38.13%) and the Western Offshore (23.33%).

Distribution of Reserves by State/Region:

The following table shows the estimated crude petroleum and natural gas reserves in India by state/region as on 31st March 2017.

Region	Crude oil reserves (in million metric tonnes)	Share of oil (%)	Natural gas reserves (in BCM)	Share of gas (%)
Arunachal Pradesh	1.52	0.25	0.93	0.07
Andhra Pradesh	8.15	1.35	48.31	3.75
Assam	159.96	26.48	158.57	12.29
Coal Bed Methane	0	0	106.58	8.26
Eastern Offshore	40.67	6.73	507.76	39.37
Gujarat	118.61	19.63	62.28	4.83
Nagaland	2.38	0.39	0.09	0.01
Rajasthan	24.55	4.06	34.86	2.70
Tamil Nadu	9.00	1.49	31.98	2.48
Tripura	0.07	0.01	36.10	2.80
Western Offshore	239.20	39.60	302.35	23.44
Total	**604.10**	**100**	**1,289.81**	**100**

Strategic Petroleum Reserves:

Building petroleum reserves like underground tank storage, above-ground tank storage, and fully developed and ready-to-exploit *in-situ* reserves is a lucrative proposition for an oil-importing country like India as the oil exporters charge exorbitant prices when the oil demand is little more than supply. The Indian Strategic Petroleum Reserve (SPR) is an emergency fuel store of total 5 million tonne or 31.5 million barrels (5.0 million cubic metres) of strategic crude oil enough to provide 10 days of consumption which are maintained by the Indian Strategic Petroleum Reserves Limited. Strategic crude oil storages are at 3 underground locations in Mangalore, Visakhapatnam and Padur with ready access to the refineries on the east and west coasts. Another method to build up strategic petroleum reserve at low cost is to develop a proven oil field for higher oil extraction rate and keeping it reserved for full production on an intermittent basis when the global oil price cross the set upper limit.

Two more SPRs will add strategic petroleum reserves of 12 days in addition to 10 days of reserves achieved in Phase I. These SPRs under Phase II will be located at Chandikhol in Odisha and Padur in Karnataka. Indian refiners maintain 65 days of crude storage, and when added to the SPR storage planned and achieved takes the Indian crude storage tally to 87 days. This is very close to the storage of 90 days mandated by IEA for member countries. The total storage figure is excluding the storage capacity of petroleum products with the marketing agencies and bulk consumers.

1.2 COMPOSITION OF PETROLEUM AND PETROLEUM PRODUCTS [W-13]

Crude oil is the term for "unprocessed" oil, the stuff that comes out of the ground. It is also known as **petroleum**. Crude oil is a **fossil fuel**, meaning that it was made naturally from decaying plants and animals living in ancient seas millions of years ago, most places you can find crude oil were once sea beds. Crude oils vary in colour, from clear to tar-black, and **in viscosity**, from water to almost solid.

Crude oils are such a useful starting point for so many different substances because they contain **hydrocarbons**. Hydrocarbons are molecules that contain hydrogen and carbon and come in various lengths and structures, from straight chains to branching chains to rings.

On average, crude oils are made of the following elements or compounds:

Carbon - 84%

Hydrogen - 14%

Sulfur - 1 to 3% (hydrogen sulfide, sulfides, disulfides, elemental sulfur)

Nitrogen - less than 1% (basic compounds with amine groups)

Oxygen - less than 1% (found in organic compounds such as carbon dioxide, phenols, ketones, carboxylic acids)

Metals - less than 1% (nickel, iron, vanadium, copper, arsenic)

Salts - less than 1% (sodium chloride, magnesium chloride, calcium chloride)

There are two things that make hydrocarbons exciting to chemists :

1. Hydrocarbons contain a lot of energy. Many of the things derived from crude oil like gasoline, diesel fuel, paraffin wax and so on take advantage of this energy.
2. Hydrocarbons can take on many different forms. The smallest hydrocarbon is **methane** (CH_4), which is a gas that is lighter than air. Longer chains with 5 or more carbons are liquids. Very long chains are solids like wax or tar. By chemically cross-linking hydrocarbon chains, you can get everything from synthetic rubber to nylon to the plastic in tupperware. Hydrocarbon chains, are very versatile.

The actual overall properties of each different petroleum source are defined by the percentage of the four main hydrocarbons found within petroleum as part of the petroleum composition.

The percentages for these hydrocarbons can vary greatly, giving the crude oil a quite distinct compound personality depending upon geographic region. These hydrocarbons are typically present in petroleum at the following percentages: paraffins (15% - 60%), napthenes (30% - 60%), aromatics (3% to 30%), with asphaltics making up the remainder.

The **composition of petroleum** is defined as laid out above, and it is this composition which gives the crude oil its properties.

Raw petroleum is usually dark brown or almost black although some fields deliver a greenish or sometimes yellow petroleum. Depending upon the field and the way the petroleum composition was formed, the crude oil will also differ in viscosity.

At the extreme ranges petroleum can be almost solid, and require a significant investment of resources to refine into a useable state as anything other than bitumen. At the other end of the scale the petroleum composition can be a clear fluid resembling kerosene or gasoline, needing very little refining to be useable as a fuel.

When discussing the composition of petroleum it is important to note that the compound of the raw petroleum tends to dictate the usage of the refined product. Petroleum is generally measured in volume, and for some composition of petroleum it is not cost effective to refine these into fuel.

A lighter, less dense raw petroleum composition with a compound that contains higher percentages of hydrocarbons is much more profitable as a fuel source. Whereas other, denser petroleum composition with a less flammable level of hydrocarbons and sulphur are expensive to refine into a fuel and are therefore more suitable for plastics manufacturing and other uses.

Unfortunately the worlds reserves of light petroleum (light crude oil) are severely depleted and refineries are forced to refine and process more and more heavy crude oil and bitumen.

In some cases, the refining process will need to remove carbon and add hydrogen, adding an extra, costly step to the refining process. This change in compound of the world's energy producing petroleum and the associated rise in refining costs has directly affected the price of gasoline across the world.

1.2.1 Hydrocarbons in Crude Oil

Petroleum consists of mixtures of hydrocarbons, which is the name given to those compounds in which the molecules consist solely of atoms of carbon and hydrogen. There exist a great variety of these hydrocarbons, depending upon the number of carbon and hydrogen atoms in each molecule and the way in which the various atoms are linked one with the other. Fortunately, as we shall see, they fall into several distinct groups or series, which helps greatly in studying them.

[a] Saturated Hydrocarbons :

The hydrocarbon methane consists of carbon with all its valencies combined with separate hydrogen atoms. This type of compound is known as a 'Saturated' compound, since there are no possibilities of further combination of the molecule with anything else. A series of these saturated hydrocarbons exists in which chains of carbon atoms are linked together with hydrogen occupying the remaining carbon valencies. At each end of the chain will be CH_3 group.

Thus, one can write :

Propane : $CH_3-CH_2-CH_3$

or Butane : $CH_3-CH_2-CH_2-CH_3$

The whole series of such hydrocarbons is known as the 'paraffin' series, and when the carbon atoms form a straight chain, so to speak (as above), the series is known as the 'normal paraffin' series.

[b] Unsaturated Hydrocarbons :

In addition to the hydrocarbons in which every valency or bond is satisfied by a separate atom, there are hydrocarbons in which carbon is linked to carbon by two or more of the valencies. In this case, the number of hydrogen atoms that can be linked to the carbon atoms is reduced. For example, in the hydrocarbon C_2H_4 or $H_2C=CH_2$ the carbon atoms are linked together by two bonds, and only two additional hydrogen atoms can then be linked to each carbon atom. It will be noticed that each carbon atom still has four valencies, but they are not all used up in separate combination with other atoms. Such a hydrocarbon is called an 'unsaturated' hydrocarbon, and the one referred to above is known as ethylene. Hydrocarbons of this class, known as 'olefins', are not found in natural crude petroleum, but they are formed in the cracking process. Being unsaturated, they can readily take part in chemical reactions and can combine with each other (polymerize), both of which facts form significant bases for some of the processes which will be discussed in later chapters.

[c] Rich Hydrocarbons :

The hydrocarbons so far discussed have been all of the 'chain' variety. There are however, other series of hydrocarbons in which carbon atoms form closed rings.

Two of the principal series found in petroleum are :

(i) The Naphthenes or Cycloparaffins : These compounds are hydrocarbons in which the carbon atoms forming a ring are fully saturated. This means that, in addition to the bonds connecting the carbon atoms with each other, the remaining two bonds on each carbon atom are linked to separate atoms of hydrogen or an equivalent group. The ring may contain various numbers of carbon atoms, though those most frequently found in petroleum contain 5, 6 or 7.

(ii) The Aromatics : The second important series of ring hydrocarbons is the series whose simple member is benzene. This is a six-carbon ring hydrocarbon in which each carbon atom in the ring has only one hydrogen atom attached.

1.2.2 Composition of Crude Oil

The crude oil mixture is composed of the following groups :

1. Hydrocarbon compounds (compounds made of carbon and hydrogen).
2. Non-hydrocarbon compounds.
3. Organometallic compounds and inorganic salts (metallic compounds).

(1) Hydrocarbon Compounds :

The principal constituents of most crude oils are hydrocarbon compounds. All hydrocarbon classes are present in the crude mixture, except alkenes and alkynes. This may indicate that crude oils originate under a reducing atmosphere.

The following is a brief description of the different hydrocarbon classes found in all crude oils.

(a) Alkanes (Paraffins) : Alkanes are saturated hydrocarbons having the general formula C_nH_{2n+2}. The simplest alkane, methane (CH_4), is the principal constituent of natural gas. Methane, ethane, propane, and butane are gaseous hydrocarbons at ambient temperatures and atmospheric pressure. They are usually found associated with crude oils in a dissolved state.

Normal alkanes (n-alkanes, n-paraffins) are straight-chain hydrocarbons having no branches. Branched alkanes are saturated hydrocarbons with an alkyl substituent or a side branch from the main chain. A branched alkane with the same number of carbons and hydrogens as an n-alkane is called an *isomer*. For example, butane (C_4H_{10}) has two isomers, n-butane and 2-methyl propane (isobutane). As the molecular weight of the hydrocarbon increases, the number of isomers also increases. Pentane (C_5C_{12}) has three isomers; hexane (C_6H_{14}) has five isomers.

The following structures shows the isomers of hexane :

$CH_3(CH_2)_4CH_3$

n-Hexane

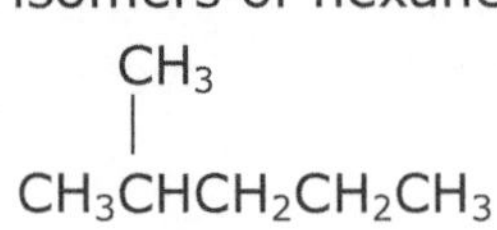

2-Methylpentane (Isohexane)

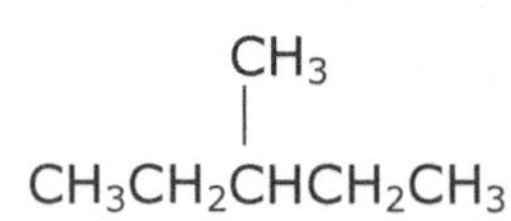

3-Methylpentane

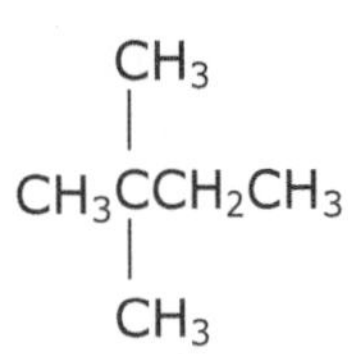

2, 2-Dimethylbutane

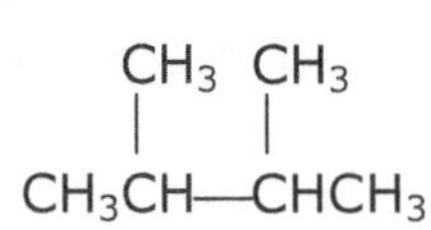

2, 3-Dimethylbutane

An iso-paraffin is an isomer having a methyl group branching from carbon number 2 of the main chain. Crude oils contain many short, medium, and long-chain normal and branched paraffins. A naphtha fraction (obtained as a light liquid stream from crude fractionation) with a narrow boiling range may contain a limited but still large number of isomers.

(b) Cycloparaffins (Naphthenes) : Saturated cyclic hydrocarbons, normally known as naphthenes, are also part of the hydrocarbon constituents of crude oils. Their ratio, however, depends on the crude type. The lower members of naphthenes are cyclopentane, cyclohexane, and their mono-substituted compounds. They are normally present in the light and the heavy naphtha fractions. Cyclohexanes, substituted cyclopentanes, and substituted cyclohexanes are important precursors for aromatic hydrocarbons.

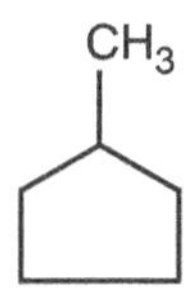

Methylcyclopentane

Cyclohexane

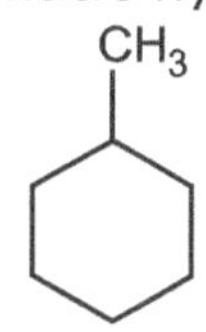

Methylcyclohexane

The examples shown here are for three naphthenes of special importance. If a naphtha fraction contains these compounds, the first two can be converted to benzene, and the last compound can dehydrogenate to toluene during processing. Dimethylcyclohexanes are also important precursors for xylenes (see "Xylenes" later in this section).

Heavier petroleum fractions such as kerosene and gas oil may contain two or more cyclohexane rings fused through two vicinal carbons.

(c) Aromatic Compounds : Lower members of aromatic compounds are present in small amounts in crude oils and light petroleum fractions. The simplest mononuclear aromatic compound is benzene (C_6H_6). Toluene (C_7H_8) and Xylene (C_8H_{10}) are also mononuclear aromatic compounds found in variable amounts in crude oils. Benzene, toluene and xylenes (BTX) are important petrochemical intermediates as well as valuable gasoline components.

Separating BTX aromatics from crude oil distillates is not feasible because they are present in low concentrations. Enriching a naphtha fraction with these aromatics is possible through a catalytic reforming process. Chapter 3 discusses catalytic reforming.

Binuclear aromatic hydrocarbons are found in heavier fractions than naphtha. Trinuclear and polynuclear aromatic hydrocarbons, in combination with heterocyclic compounds, are major constituents of heavy crudes and crude residues. Asphaltenes are a complex mixture of aromatic and heterocyclic compounds. The nature and structure of some of these compounds have been investigated.

The following are representative examples of some aromatic compounds found in crude oils :

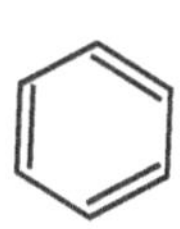

Benzene

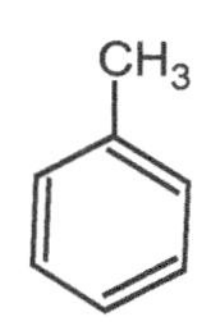

Toluene

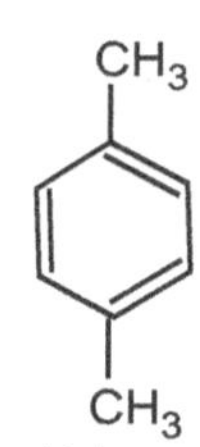

p - Xylene

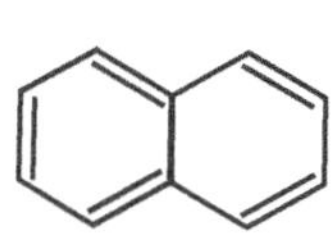

Naphthalene

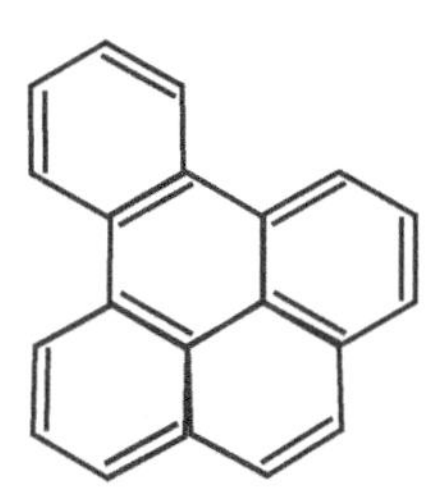

1,2 - Benzopyrene

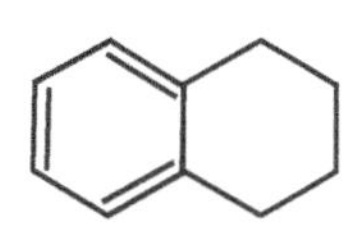

Tetralin

Only a few aromatic-cycloparaffin compounds have been isolated and identified. Tetralin is an example of this class.

(2) Non-hydrocarbon Compounds :

Various types of non-hydrocarbon compounds occur in crude oils and refinery streams. The most important are the organic sulfur, nitrogen and oxygen compounds. Traces of metallic compounds are also found in all crudes. The presence of these impurities is harmful and may cause problems to certain catalytic processes. Fuels having high sulfur and nitrogen levels cause pollution problems in addition to the corrosive nature of their oxidization products.

(a) Sulfur Compounds : Sulfur in crude oils is mainly present in the form of organosulfur compounds. Hydrogen sulfide is the only important inorganic sulfur compound found in crude oil. Its presence, however, is harmful because of its corrosive nature. Organosulfur compounds may generally be classified as acidic and non-acidic. Acidic sulfur compounds are the thiols (mercaptans). Thiophene, sulfides and disulfides are examples of non-acidic sulphur compounds found in crude fractions. Extensive research has been carried out to identify some sulfur compounds in a narrow light petroleum fraction.

Examples of some sulfur compounds from the two types are:

(i) Acidic Sulphur Compound :

CH_3SH

Methyl mercaptan

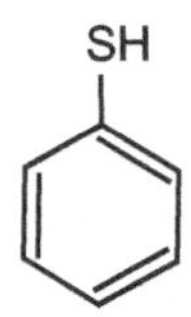

Phenyl mercaptan

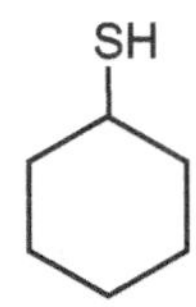

Cyclohexylthiol

(ii) Non-Acidic Sulphur Compound :

CH_3SCH_3

Dimethyl sulphide

$CH_3S\text{-}SCH_3$

Dimethyldisulfide

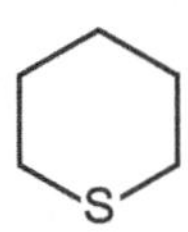

Thiocyclohexane

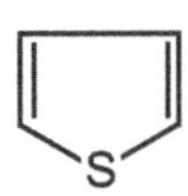

Thiophene

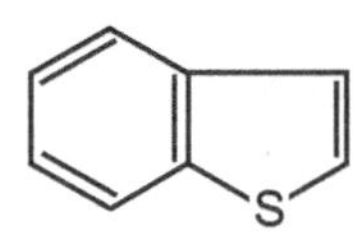

Benzothiophene

Sour crudes contain a high percentage of hydrogen sulfide. Because many organic sulfur compounds are not thermally stable, hydrogen sulphide is often produced during crude processing. High-sulfur crudes are less desirable because treating the different refinery streams for acidic hydrogen sulfide increases production costs.

Most sulfur compounds can be removed from petroleum streams through hydrotreatment processes, where hydrogen sulfide is produced and the corresponding hydrocarbon is released. Hydrogen sulfide is then absorbed in a suitable absorbent and recovered as sulfur.

(b) Nitrogen Compounds : Organic nitrogen compounds occur in crude oils either in a simple heterocyclic form as in pyridine (C_5H_5N) and pyrrole (C_4H_5N), or in a complex structure as in porphyrin. The nitrogen content in most crudes is very low and does not exceed 0.1 wt%. In some heavy crudes, however, the nitrogen content may reach upto 0.9 wt%. Nitrogen compounds are more thermally stable than sulfur compounds and accordingly are concentrated in heavier petroleum fractions and residues. Light petroleum streams may contain trace amounts of nitrogen compounds, which should be removed because they poison many processing catalysts. During hydrotreatment of petroleum fractions, nitrogen compounds are hydrodenitrogenated to ammonia and the corresponding hydrocarbon.

For example, pyridine is denitrogenated to ammonia and pentane.

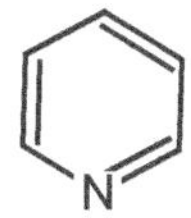 $+ 5H_2 \longrightarrow NH_3 + CH_3CH_2CH_2CH_2CH_3$

Nitrogen compounds in crudes may generally be classified into basic and non-basic categories. Basic nitrogen compounds are mainly those having a pyridine ring, and the non-basic compounds have a pyrrole structure. Both pyridine and pyrrole are stable compounds due to their aromatic nature.

The following are examples of organic nitrogen compounds :

(i) Basic Nitrogen Compound :

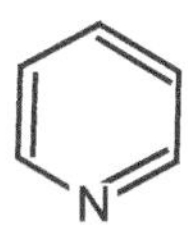

Pyridine

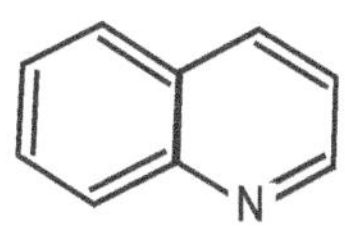

Quinoline

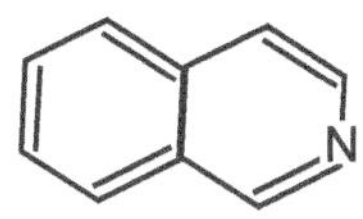

Isoquinoline

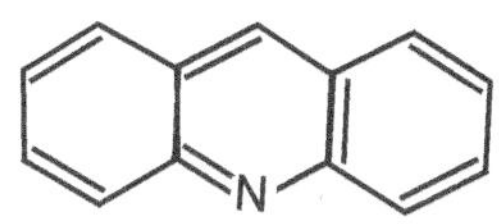

Acridine

(ii) Non-Basic Nitrogen Compound :

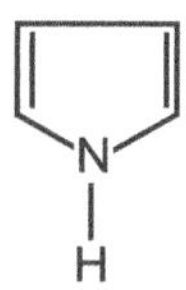

Pyrrole

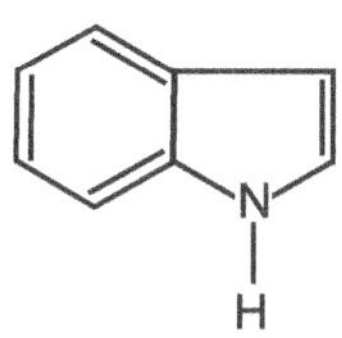

Indole

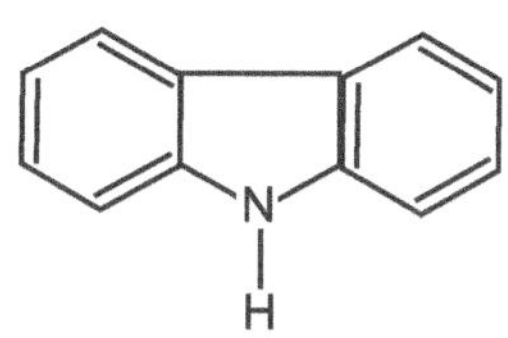

Carbazole

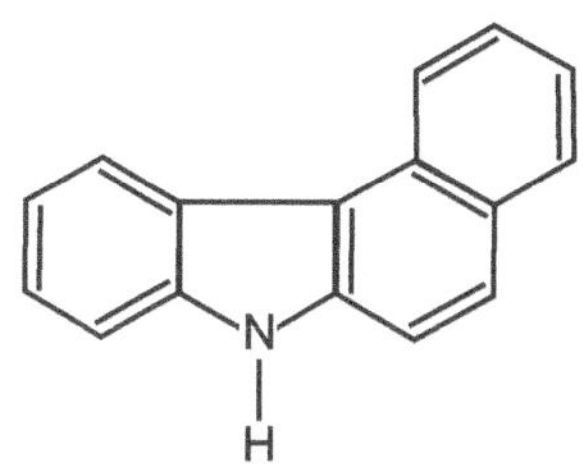

Benzocarbazole

Porphyrins are non-basic nitrogen compounds. The porphyrin ring system is composed of four pyrrole rings joined by =CH-groups. The entire ring system is aromatic. Many metal ions can replace the pyrrole hydrogens and form chelates. The chelate is planar around the metal ion and resonance results in four equivalent bonds from the nitrogen atoms to the metal. Almost all crude oils and bitumens contain detectable amounts of vanadyl and nickel porphyrin.

The structure of Porphyry is shown below:

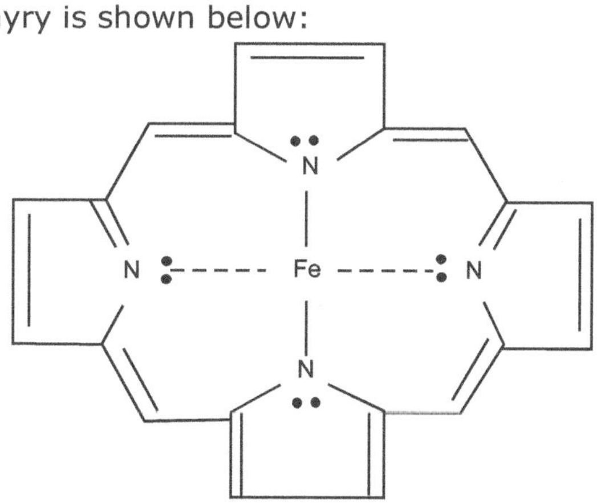

Porphyrin structure

Separation of nitrogen compounds is difficult, and the compounds are susceptible to alteration and loss during handling. However, the basic low molecular weight compounds may be extracted with dilute mineral acids.

(c) Oxygen Compounds : Oxygen compounds in crude oils are more complex than the sulphur types. However, their presence in petroleum streams is not poisonous to processing catalysts. Many of the oxygen compounds found in crude oils are weakly acidic.

They are carboxylic acids, cresylic acid, phenol and naphthenic acid. Naphthenic acids are mainly cyclopentane and cyclohexane derivatives having a carboxyalkyl side chain.

Naphthenic acids in the naphtha fraction have a special commercial importance and can be extracted by using dilute caustic solutions. The total acid content of most crudes is generally low, but may reach as much as 3%, as in some California crudes.

Non-acidic oxygen compounds such as esters, ketones and amides are less abundant than acidic compounds. They are of no commercial value.

Below are given some of the oxygen compounds commonly found in crude oils :

(i) Acidic Oxygen Compounds :

$CH_3(CH_2)_nCOOH$

Analiphatic carboxylic acid

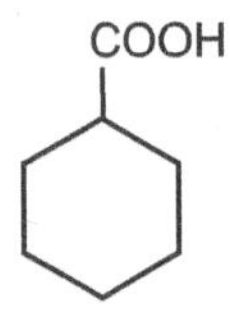

Cyclohexane carboxylic acid

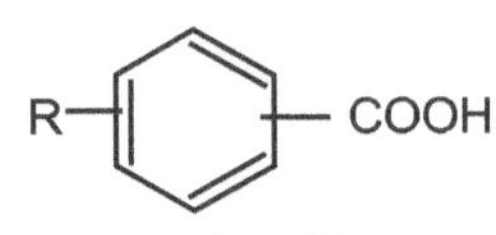

Aromatic acids

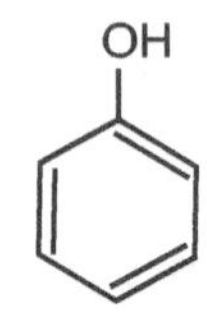

Phenol

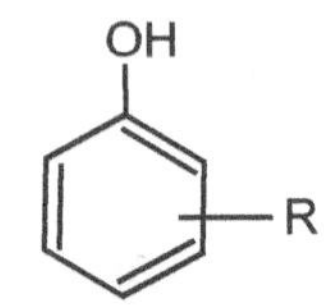

Cresylic acid

(ii) Non-Acidic Oxygen Compounds :

R-COOR'

Esters

R-CONHR'

Amides

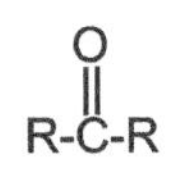

Ketone

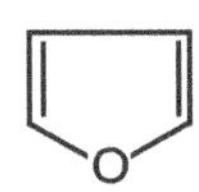

Furan

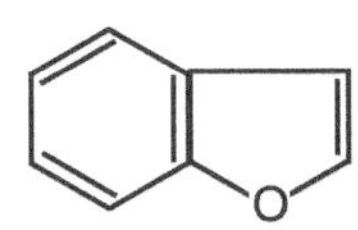

Benzofuran

(3) Metallic Compounds :

Many metals occur in crude oils. Some of the more abundant are sodium, calcium, magnesium, aluminium, iron, vanadium and nickel. They are present either as inorganic salts, such as sodium and magnesium chlorides, or in the form of organometallic compounds, such as those of nickel and vanadium (as in porphyrins). Calcium and magnesium can form salts or soaps with carboxylic acids. These compounds act as emulsifiers, and their presence is undesirable.

Although metals in crudes are found in trace amounts, their presence is harmful and should be removed. When crude oil is processed, sodium and magnesium chlorides produce hydrochloric acid, which is very corrosive. Desalting crude oils is a necessary step to reduce these salts. Vanadium and nickel are poisons to many catalysts and should be reduced to very low levels. Most of the vanadium and nickel compounds are concentrated in the heavy residues. Solvent extraction processes are used to reduce the concentration of heavy metals in petroleum residues.

1.3 CLASSIFICATION OF PETROLEUM

For several decades now the crude oil or petroleum industry has classified the raw crude by the location from which it was extracted. In other words, oil is classified by geographic region. However, all oils from a particular region are not same. Further classification of petroleum, derived from the density of the raw petroleum (API gravity) and its various non-hydrocarbon components (especially sulfur), is then added to the geographic designation. The end result of all this classification helps determine the price of a specific barrel of crude as well as how much demand there is for that particular oil.

In general, if the crude oil contains high levels of sulphur, the petroleum classification is termed 'sour, if it has relatively low levels of sulphur the petroleum classification is termed 'sweet'. If the raw petroleum is of a high density then the petroleum classification is termed 'heavy' and if it is of a low density the petroleum classification is termed 'light'. Density of oil is determined by the length of the hydrocarbons it contains. If it contains a great deal of long-chain hydrocarbons, the petroleum will be denser. If it contains a greater proportion of short-chain hydrocarbons it will be less dense. Besides chain length, the ratio of carbon to hydrogen also helps to determine the density of a particular hydrocarbon. The greater the amount of hydrogen in relation to carbon, the lighter the hydrocarbon and will be less dense oil when will float on top of denser oil and is generally easier to pump.

The hydrocarbons in crude oil can generally be divided into four categories:

- **Paraffins:** These can make up 15 to 60% of crude and have a carbon to hydrogen ratio of 1:2, which means they contain twice the amount of hydrogen as they do carbon. These are generally straight or branched chains, but never cyclic (circular) compounds. Paraffins are the desired content in crude which are used to make fuels. The shorter the paraffins are, the lighter the crude is.
    - General formula: C_nH_{2n+2} (n is a whole number, usually from 1 to 20).
    - Straight or branched-chain molecules can be gasses or liquids at room temperature depending upon the molecule.
    - Examples : methane, ethane, propane, butane, isobutane, pentane, hexane.
- **Napthenes or Cycloalkanes:** These can make up 30 to 60% of crude and have a carbon to hydrogen ratio of 1:2. These are cyclic compounds and can be thought of as cycloparaffins. They are higher in density than equivalent paraffins and are more viscous.
    - General formula : C_nH_{2n} (n is a whole number usually from 1 to 20).
    - Ringed structures with one or more rings.
    - Rings contain only single bonds between the carbon atoms typically liquids at room temperature.
    - Examples : cyclohexane, methyl cyclopentane.
- **Aromatics:** These can constitute anywhere from 3 to 30% of crude. They are undesirable because burning them results in soot. They have a much less hydrogen in comparison to carbon than is found in paraffins. They are also more viscous. They are often solid or semi-solid when an equivalent paraffin would be a viscous liquid under the same conditions.
    - General formula: C_6H_{5-Y} (Y is a longer, straight molecule that connects to the benzene ring).
    - Ringed structures with one or more rings.
    - Rings contain six carbon atoms, with alternating double and single bonds between the carbons, typically liquids.
    - Examples : benzene, naphthalene.
- **Asphaltics:** These have average of about 6% in most crudes. They have a carbon to hydrogen ratio of approximately 1:1, making them very dense. They are generally undesirable in crude, but their 'stickiness' makes them excellent for use in road construction.
- **Other Hydrocarbons :**
    - **Alkenes :**
        - General formula: C_nH_{2n} (n is a whole number, usually from 1 to 20).
        - Linear or branched chain molecules containing one carbon-carbon double-bond can be liquid or gas.
        - Examples : ethylene, butene, isobutene.

- **Dienes and Alkynes :**
  - General formula: C_nH_{2n-2} (n is a whole number, usually from 1 to 20).
  - Linear or branched chain molecules containing two carbon-carbon double-bonds can be liquid or gas.
  - Examples : acetylene, butadienes.

When considering the **petroleum classification**, it is important to consider the fact that the overall classification will have an effect on the value, not just on the physical properties. For example, petroleum with a geographical classification from one region of the world may be expensive to transport to another region of the world regardless of the suitability of the raw petroleum as an overall substance. In general, lighter crude commands a higher price because it contains more hydrocarbon chains that can be easily refined to make gasoline and diesel, which are in high demand. The lower the sulphur content, the higher the price as well because low-sulfur, sweet crude requires less refining.

Classification of petroleum also indicates the best use for a particular field of petroleum. One oil type is not necessarily "better" than another, but rather the different types are useful in different applications. Light crude oil is preferable for refining into gasoline as it produces a far higher yield than heavy. In a similar fashion, sweet petroleum is often more desirable than sour petroleum as its use will cause far less impact on the environment in the form of harmful emissions as it is burned. These basic classifications of petroleum are further enhanced by a full molecular description gained through a crude oil assay analysis.

1.4 INDIAN PETROLEUM REFINERY AND DIFFERENT SECTION AND EQUIPMENT

1.4.1 Production of Crude Oil in India (S-14)

India is not among the major producers of crude oil, as it does not have much oil reserves. That is why, it generally depends on imports of crude oil from other countries. However, the production of oil and as a result the production of its by-products in India has increased in the recent past due to exploration and findings of new oil reserves. India currently has an estimated quantity of 5.4 billion barrels of oil reserves, out of which it produces around 0.8 million barrels per day. At this production level, the oil reserves in India would last for around 29 years.

The major oil reserves of the country are situated at :

- Mumbai high (Mumbai)
- Upper Assam (Assam)
- Cambay (Gujarat)
- Krishna-Godavari basin (Andhara Pradesh)
- Cauvery basin (Tamil Nadu)
- Nagaland
- Arunachal Pradesh

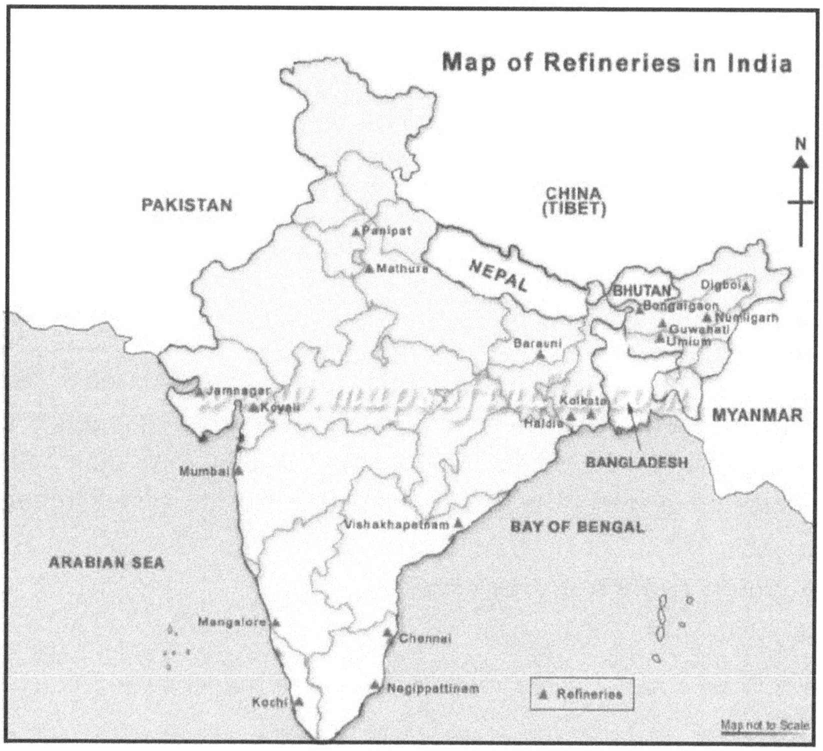

Fig. 1.1 : Map of refineries in India

The largest crude oil producing oilfield is the **Mumbai high** field that produces around **2,60,000 barrels per day**. Among these production centers, major share of production i.e. **2/3rd share** is bagged by the offshore reserves as compared to onshore reserves. The refining capacity of crude oil in India is over 2.1 million barrels per day. The refining sector in India is held by both public and private sector, public sector being the dominating one.

1.4.2 Indian Crude Oil Market (S-16, 17, 18; W-17)

India is one of the non-OPEC countries much dependent on its imports to fulfill the domestic consumption demand as it has a much lower level of production. India is a developing country and the requirement for the oil as a primary energy constituent from the industries in the country is at its peak. The country has much depend on coal to satisfy its energy needs in the earlier times but the use of crude oil and gas is taking over the dominance of coal with the change in time. Oil and gas contribute to around 45% of the country's total energy consumption.

India has around 5.4 billion barrels of oil reserves with it and the domestic production has increased in the recent past to reach the 0.8 million barrels per day mark. Mumbai high is the largest oil-producing oilfield in India with a production of 2.6 lakh barrels per day. The refining capacity of crude oil in India is estimated at around 2.1 million barrels per day. Regarding the consumption pattern of oil in India, it is the 6th largest consumer country in the world having a consumption of 2.2 million barrels per day. This leaves the country with a huge deficit in the demand-supply scenario and thus 70% of the consumption is met through imports.

Table 1.1: Major crude oil producers

Sr. No.	Name of company with location	Production capacity
1.	Reliance, Hazira	1.6 MMTPA
2.	IPCL, Baroda	1.0 MMTPA
3.	HPCL, Mumbai	0.43 MMTPA
4.	GAIL, Aurya	0.30 MMTPA
5.	NOCIL, Thane	0.30 MMTPA
6.	Others	0.31 MMTPA

India generally imports Oman-Dubai sour grade crude, Brent dated sweet crude and Bonny light crude. The country imports over 1.5 million barrels per day that place it at the 9th position among the largest importers of the world. Though the Indian production has increased in the recent times, the imports were raised by 5% making due to the raised Indian demand of around 4.2%.

The countries from which India imports crude oil are :

- Venezuela
- Nigeria
- Sudan
- Iran
- Kuwait

The Indian oil-refining sector has been regulated by the government historically and is still dominated. A new private sector has emerged after the loosening of control by the government.

Major Units Pertaining to the Oil Sector in India :

- Indian Oil Corporation (Public sector)
- Oil and Natural Gas Corporation (Public sector)
- Reliance India Ltd. (Private sector)
- Essar Oil Refinery (Private sector)
- Bharat Petroleum Corporation Ltd. (Public sector)
- Hindustan Petroleum Corporation Ltd. (Public sector)
- Mangalore Refineries and Petrochemicals Ltd. (Public sector)

Market Influencing Factors :

- Production of the major oil producing countries.
- Various climatic or political supply fluctuations.
- World oil demand.
- Fluctuations in the value of dollar.
- Imports from various world oil organizations like API, DOE.
- Refinery Fire.

Major Trading Centers of Crude Oil :

The major trading centers of crude oil in the world are :

- New York Mercantile Exchange (NYMEX).
- International Petroleum Exchange of London (IPE).
- Tokyo Commodity Exchange (TOCOM).

Table 1.2: Indian oil refineries: Installed capacities (MMTPA)

Sr. No.	Refinery plants	Capacity (MMPTA)
1.	Digboi	0.65
2.	Guwahati	1.0
3.	Barauni	6.0
4.	Koyali	13.7
5.	Haldia	6.0
6.	Mathura	8.0
7.	Panipat	12.0
	Subtotal	**47.35**
8.	CPCL - Chennai	9.5
9.	Narimanam	1.0
10.	Bongaigaon	2.35
	Subtotal	**12.85**
	Group Total	**60.2**

(MMTPA – Million Metric Tonnes Per Annum, equal to 20, 000 barrels per day).

Table 1.3: Production of Petroleum (Crude) in India

State/Area	Production (tonnes)	Percentage of all India production	Value (₹ crore)
1. Mumbai High	21,573	65.28	12,016.16
2. Gujarat	6,042	18.28	3,365.94
3. Assam	4,659	14.09	2,395.6
4. Tamil Nadu	395	1.19	220.01
5. Andhra Pradesh	300	0.93	167.10
6. Arunachal Pradesh	74	0.23	41.22
All India	**33,043**	**100.00**	**18,404.95**

1.5 REFINERY PRODUCTS

An **oil refinery** or **petroleum refinery** is an industrial process plant where crude oil is transformed and refined into more useful products such as petroleum naphtha, gasoline, diesel fuel, asphalt base, heating oil, kerosene, liquefied petroleum gas, jet fuel and fuel oils. Petrochemicals feed stock like ethylene and propylene can also be produced directly by cracking crude oil without the need of using refined products of crude oil such as naphtha.

Oil refineries are typically large, sprawling industrial complexes with extensive piping running throughout, carrying streams of fluids between large chemical processing units, such as distillation columns. In many ways, oil refineries use much of the technology, and can be thought of, as types of chemical plants.

The crude oil feedstock has typically been processed by an oil production plant. There is usually an oil depot at or near an oil refinery for the storage of incoming crude oil feedstock as well as bulk liquid products.

Petroleum refineries are very large industrial complexes that involve many different processing units and auxiliary facilities such as utility units and storage tanks. Each refinery has its own unique arrangement and combination of refining processes largely determined by the refinery location, desired products and economic considerations.

An oil refinery is considered an essential part of the downstream side of the petroleum industry.

Some modern petroleum refineries process as much as 800,000 to 900,000 barrels (127,000 to 143,000 cubic metres) of crude oil per day.

According to the Oil and Gas Journal in the world a total of 636 refineries were operated on the 31st December 2014 for a total capacity of 87.75 million barrels (13,951,000 m^3).

Jamnagar Refinery is the largest oil refinery, since 25th December 2008, with a processing capacity of 1.24 million barrels (197,000 m^3). Located in Gujarat, India, it is owned by Reliance Industries.

Major Products:

Petroleum products are materials derived from crude oil (petroleum) as it is processed in oil refineries. The majority of petroleum is converted to petroleum products, which includes several classes of fuels.

Oil refineries also produce various intermediate products such as hydrogen, light hydrocarbons, reformate and pyrolysis gasoline. These are not usually transported but instead are blended or processed further on-site. Chemical plants are thus often adjacent to oil refineries or a number of further chemical processes are integrated into it. For example, light hydrocarbons are steam-cracked in an ethylene plant, and the produced ethylene is polymerized to produce polyethene.

Because of technical reasons and environment protection demand a very low sulfur content in all but the heaviest products, is transformed to hydrogen sulfide via catalytic hydrodesulfurization and removed from the product stream via amine gas treating. Using the Claus process, hydrogen sulfide is afterwards transformed to elementary sulfur to be sold to the chemical industry. The rather large heat energy freed by this process is directly used in the other parts of the refinery. Often an electrical power plant is combined into the whole refinery process to take up the excess heat.

According to the composition of the crude oil and depending on the demands of the market, refineries can produce different shares of petroleum products. The largest share of oil products is used as "energy carriers", i.e. various grades of fuel oil and gasoline. These fuels include or can be blended to give gasoline, jet fuel, diesel fuel, heating oil, and heavier fuel oils. Heavier (less volatile) fractions can also be used to produce asphalt, tar, paraffin wax, lubricating and other heavy oils. Refineries also produce other chemicals, some of which are used in chemical processes to produce plastics and other useful materials. Since petroleum often contains a few percent sulfur-containing molecules, elemental sulfur is also often produced as a petroleum product. Carbon, in the form of petroleum coke, and hydrogen may also be produced as petroleum products. The hydrogen produced is often used as an intermediate product for other oil refinery processes such as hydrocracking and hydrodesulfurization.

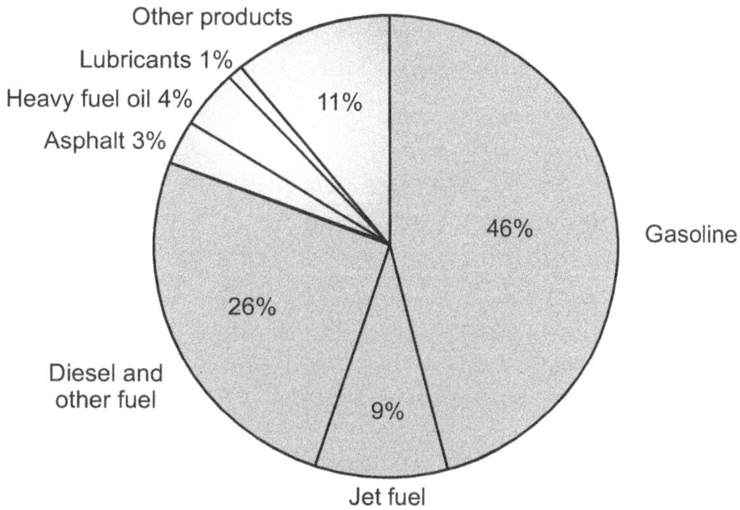

Fig 1.2: A breakdown of the products made from a typical barrel of oil

Petroleum products are usually grouped into four categories: light distillates (LPG, gasoline, naphtha), middle distillates (kerosene, jet fuel, diesel), heavy distillates and residuum (heavy fuel oil, lubricating oils, wax, asphalt). These require blending various feedstocks, mixing appropriate additives, providing short term storage, and preparation for bulk loading to trucks, barges, product ships and railcars. This classification is based on the way crude oil is distilled and separated into fractions.

- Gaseous fuel such as Liquified petroleum gas and propane is stored and shipped in liquid form under pressure.
- Lubricants (produces light machine oils, motor oils, and greases, adding viscosity stabilizers as required), usually shipped in bulk to an offsite packaging plant.
- Paraffin wax, used in the packaging of frozen foods, among others. May be shipped in bulk to a site to prepare as packaged blocks. Used for wax emulsions, construction board, matches, candles, rust protection, and vapor barriers.
- Sulfur (or sulfuric acid), byproducts of sulfur removal from petroleum which may have up to a couple percent sulfur as organic sulfur-containing compounds. Sulfur and sulfuric acid are useful industrial materials. Sulfuric acid is usually prepared and shipped as the acid precursor oleum.
- Bulk tar shipping for offsite unit packaging for use in tar-and-gravel roofing.
- Asphalt used as a binder for gravel to form asphalt concrete, which is used for paving roads, lots etc. An asphalt unit prepares bulk asphalt for shipment.
- Petroleum coke, used in speciality carbon products like electrodes or as solid fuel.
- Petrochemicals are organic compounds that are the ingredients for the chemical industry, ranging from polymers and pharmaceuticals, including ethylene and benzene-toluene-xylenes ("BTX") which are often sent to petrochemical plants for further processing in a variety of ways. The petrochemicals may be olefins or their precursors, or various types of aromatic petrochemicals.
    - Gasoline
    - Naphtha
    - Kerosene and related jet aircraft fuels
    - Diesel fuel and Fuel oils
    - Heat
    - Electricity

Over 6,000 items are made from petroleum waste by-products including: fertilizer, floor coverings, perfume, insecticide, petroleum jelly, soap, vitamin capsules.

1.5.1 Modern Refining

Petroleum refineries are marvels of modern engineering. Within them a maze of pipes, distillation columns, and chemical reactors turn crude oil into valuable products. Large refineries cost billions of dollars, employ several thousand workers, operate around the clock, and occupy the same area as several hundred football stadiums. The U.S. has about 300 refineries that can process anywhere between 40 and 400,000 barrels of oil a day. These refineries turn out the gasoline and chemical feedstocks that keep the country running.

The Search :

Locating an oil field is the first obstacle to be overcome. The first explorers used Y-shaped devining rods and other supernatural, but ineffective, means of locating petroleum. Today geologists and petroleum engineers employ more tried and true methods. Instruments to aid the search include; geophones (uses sound), gravimeters (uses gravity), and magnetometers (uses the Earth's magnet field). While these methods narrow the search tremendously, a person still has to drill a exploratory well, or wildcat well, to see if the oil actually exists. Success brings visions of gushers soaring skyward, however today wells are capped before this happens.

Drilling :

There are three main types of drilling operations; cable-tool, rotary and off-shore. Cable-tool drilling involves a jack-hammer approach where a chisel dislodges earth and hauls up the loose sediment. Rotary drilling works at much greater depths, and involve sinking a drill pipe with a rotating steel bit in the middle. Off-shore drilling involves huge semisubmersible platforms which lower a shaft to the ocean floor, containing any oil which is located.

All crude oil contains some amount of methane or other gases dissolved in it. Once the drilling shaft makes contact with the oil it releases the pressure in the underground reservoir. Just like opening a can of soda pop, the dissolved gases fizz out of solution pushing crude oil to the surface. The dissolved gases will allow about 20% recovery of oil. To get better recovery, water is often pumped into the well, this forces the lighter oil to the surface. Water flooding allows recoveries of about 50%. The addition of surfactant allows even more oil to be recovered by preventing much of it from getting trapped in nooks and crannies. Yet, it is impossible to get all of the oil out of a well.

Transportation :

Crude oil is a liquid, so it is much easier to move than natural gas or coal. Coal is nice and dense, so it does not require large holding containers, but it cannot be pumped. Conveyor belts and cranes cannot compete with pipelines for economic efficiency. Natural gas can be pumped using expensive compressors, but it requires enormous holding tanks. A recent trick has been to inject huge amounts of water into salt strata. The water dissolves the salt, leaving truly enormous caverns. The natural gas is then pumped in and stored until needed. The ease in transporting oil is one of the reasons we have become so dependent upon it. Pound for pound natural gas and coal just cannot compete.

Reserves :

The proven reserves of crude oil within the U.S. are about 3.9 billion cubic metres. This could cover the state of Minnesota with a layer one half inch thick. A reasonable value for the total amount of crude oil obtainable using current methods from around the world is 350 billion cubic metres. This could cover Minnesota with a layer of oil four and a half feet thick. Yet, at the rate we are consuming oil, the nation's reserves will be depleted by 2020, and the world's reserves will be depleted by the end of the 21st Century.

1.5.2 Factors Affecting the Price of Crude Oil **(S-14, 16, 18; W-13, 17)**

The crude oil prices have been buffeted by many factors, which are summarized below :

- **Production :** The OPEC nations are the major producer of world's crude oil. Therefore, every policy made by such countries related to the crude oil prices have their influence on crude oil prices. Any decision taken by OPEC nations for increasing or decreasing production of crude oil impacts the price level of crude oil in international commodity markets.
- **Natural Causes :** In recent years, global community have witnessed many events which in turns have volatility effects on the price level of crude oil. Hurricane, Katrina and other type of tropical cyclone have hit the major portion of globe, which as a result driven the crude oil prices to reach at its peak.
- **Inventory :** In throughout the world, oil producers and consumers get stock their crude oil for their future requirements. This gives rise to speculation on price expectations and sale/arbitrage chances in case any unexpected thing cracks during supply and demand equations. Any upward or downward movement in inventory level shoots up volatility in price index of crude oil, which generates lot of changing movement in sensex.
- **Demand :** With a sharp rise in economic demand, requirement of crude oil is increasing to manifold in context to the limited supply. The high demand economies of crude oil are putting undue pressure on the available fixed resources. The major gap created between demand and supply of crude oil is forcing the price curve of crude oil to rise in upward direction.

The price structure of crude oil is also influenced by the cyclical pattern. It has been observed that requirement of crude oil got increased during summer season as compared to the winter season. As any dip in the seasonal temperature increases the consumption of energy for heating purpose in many cold nations, demand shoots up and thus generates the requirement of tapping the inventories. Similarly, in summer, supply exceeds the demand and petroleum inventories are build up for storage purpose. Henceforth, crude oil prices drop.

Important Points

- "Crude oil is a yellowish black mineral oil that is extracted from under the surface of the earth". This oil consists of a number of hydrocarbon compounds as it is formed from the million of year's old animal and plant remnants.
- Crude oil is usually a dark coloured or black liquid with a characteristic odour imparted by small quantities of compounds such as those containing sulphur or nitrogen.
- Although India has vast areas covered by sedimentary rocks, structures containing oil are not in proportion to the expanses of these rocks and are found in limited situations.
- The Indian Mineral Year Book 1982 estimated a reserve of 468 million tonnes of which 328 million tonnes was available in Mumbai High. In 1984, the reserves were estimated at 500 million tonnes.
- A refinery is composed of a group of chemical engineering unit processes and unit operations used for refining certain materials or converting raw material into products of value.
- India was not known to the world in the context of crude oil and it's by-product production. As late as in 1889, the presence of oil in India was discovered in Digboi in Assam. First crude oil refinery in India was set up in Digboi in 1901.

- The various types of crude oils are classified according to their geographical originations, sulfur level and also the density of the oils.
- Mumbai high is the largest crude oil producing oilfield near around 2,60,000 barrels per day.
- India is one of the **non-OPEC** countries.
- OPEC stands for **'Organization of Petroleum Exporting Countries'** and the members include all the 11 major crude oil producing countries and nations that are highly dependent on the revenues from oil and oil products.
- The major units pertaining to the oil sector in India are: Indian Oil Corporation, Oil and Natural Gas Corporation, Reliance India Ltd., Essar Oil Refinery, Bharat Petroleum Corporation Ltd, Hindustan Petroleum Corporation Ltd, Mangalore Refineries and Petrochemicals Ltd.
- The major classes of **hydrocarbons in crude oils** are :
  1. **Paraffins :** General formula is C_nH_{2n+2}.
  2. **Aromatics :** General formula is C_6H_{5-Y}.
  3. **Alkenes :** General formula is C_nH_{2n}.
  4. **Dienes and Alkynes :** General formula is C_nH_{2n-2}.
- **Factors affecting the price of Crude oil :**
  - Production
  - Natural Causes
  - Demand
  - Inventory
- **Hydrocarbons in Crude Oil :**

  It consists of mixtures of hydrocarbons, which is the name given to those compounds in which the molecules consist solely atoms of carbon and hydrogen.
  1. **Saturated Hydrocarbons :** It consists of carbon with all its valencies combined with separate hydrogen atoms.
  2. **Unsaturated Hydrocarbons :** These are hydrocarbons in which carbon is linked to carbon by two or more of the valencies. In this case, the number of hydrogen atoms that can be linked to the carbon atoms is reduced.
  3. **Ring Hydrocarbons :** Two of the principal series found in petroleum are–

  (i) The Naphthenes or Cycloparaffins: the bonds connecting the carbon atoms with each other.

  (ii) The Aromatics: It is a six-carbon ring hydrocarbon.
- **Composition of crude oil:** It is a mixture of following groups.
  1. **Hydrocarbon Compounds :** These compounds are made up of Carbon and Hydrogen and they are classified in following groups :

     (a) Alkanes : These are straight-chain hydrocarbons having no branches. For example, CH_4, C_4H_{10} etc.

     (b) Cycloparaffins (Naphthenes) : They contain the saturated cyclic hydrocarbons ring and they are present in Naphtha fractions.

     (b) Aromatic Compounds : They are present in small amounts in crude oils and light petroleum fractions. It is the series whose simple member is benzene, Toluene and xylene.

2. **Non-hydrocarbon Compounds :** In the presence of these harmful impurities, oil may cause problems of pollution and the compounds are organic sulfur, nitrogen, and oxygen compounds.
3. **Metallic Compounds :** Presence of the metals is very harmful and therefore it should be removed. Solvent extraction processes are used to reduce the concentration of heavy metals in petroleum residues.

- The hydrocarbons in crude oil can generally be divided into four categories:
  1. Paraffins
  2. Napthenes or Cycloalkanes
  3. Aromatics
  4. Asphaltics
- Major Crude Oil Producers in India :

Sr. No.	Name of company and location	Production capacity (MMTPA)
1.	Reliance, Hazira	1.6
2.	IPCL, Baroda	1.0
3.	HPCL, Mumbai	0.43
4.	GAIL, Aurya	0.30
5.	NOCIL, Thane	0.30
6.	Others	0.31

- An **oil refinery** or **petroleum refinery** is an industrial process plant where crude oil is transformed and refined into more useful products such as petroleum naphtha, gasoline, diesel fuel, asphalt base, heating oil, kerosene, liquefied petroleum gas, jet fuel and fuel oils. Petrochemicals feed stock like ethylene and propylene can also be produced directly by cracking crude oil without the need of using refined products of crude oil such as naphtha.
- The crude oil prices have been buffeted by Production, Natural Causes, Inventory and Demand.

Practice Questions

1. Crude oil is called "Black gold". Give reason.
2. Give the chemical composition or different constituents of crude petroleum ?
3. List the names of six major oil producing countries in the world with their percentage share in world crude oil production ?

MSBTE Questions and Answers (As Per G-Scheme)

Summer 2016

1. List the name of at least six major refineries in India. Which is the largest refinery among them? **(4 M)**

Ans. Refer Section 1.4.2.

2. What is OPEC. List the names of four oil producing countries in the world. **(4 M)**

Ans. Refer Section 1.4.2.

3. Crude oil is often called black gold. Explain why? (4 M)

Ans. Refer Section 1.1.

4. What are the different factors affecting the price of crude oil? (4 M)

Ans. Refer Section 1.5.2.

5. What is difference between petroleum refinery and petrochemical industry? (4 M)

Ans. Refer Sections 1.1 and 1.2.

6. State the constituents of crude oil. List the names of four unit operations involved in refining process. (4 M)

Ans. Refer Section 1.2

Summer 2017

1. Name any four Indian refineries with their location and capacity. (4 M)

Ans. Refer Section 1.4.2.

2. Name any eight global crude oil producers. (4 M)

Ans. Refer Section 1.4.

3. Describe composition of petroleum in detail. (4 M)

Ans. Refer Section 1.6.

Winter 2017

1. What is OPEC? List the names of six major oil producing countries in the world with their percentage share in world crude oil production. (4 M)

Ans. Refer Section 1.4.2.

2. What is the difference between petroleum refineries and petrochemical industries? Name any two petroleum refinery and petrochemical industry. (4 M)

Ans. Refer 1.4.2.

3. What are the different factors affecting the price of crude oil ? (4 M)

Ans. Refer Section 1.5.2.

4. Why crude oil is called Black Gold? What are its advantages and disadvantages over other energy sources? (4 M)

Ans. Refer Section 1.1.

5. Give the classification of crude. (4 M)

Ans.: Refer Section 1.6.

Summer 2018

1. List the name of OPEC countries with their percentage crude oil production. (4 M)

Ans. Refer Section 1.4.2.

2. List the name of any four Indian refineries with their capacity. (4 M)

Ans. Refer Section 1.4.2.

3. State the constituents and characteristics of crude oil. (4 M)

Ans. Refer Section 1.2.

4. Crude oil is often called as black gold. Give reason. (4 M)

Ans. Refer Section 1.1.

5. Explain the :

   (i) Factors affecting the price of crude oil (four points). (4 M)

Ans. Refer Section 1.5.2.

Chapter **2...**

DISTILLATION OF CRUDE OIL

Weightage of Marks = 18, Teaching Hours = 12

Syllabus

2.1 Primary treatment of crude: Dehydration and desalting of crude.

2.2 Atmospheric distillation of crude.

2.3 Vacuum distillation of crude.

2.4 Fractions obtained from crude oil with their boiling range and uses.

2.5 Properties of petroleum products: Octane number, cetane number, flash point, fire point, aniline point, pour point, cloud point, drop point, calorific value, carbon residue.

2.6 Pensky Marten's apparatus, Abel's apparatus, cloud and pour point apparatus, drop point apparatus, smoke point apparatus, aniline point apparatus, Conradson apparatus.

2.7 Measurement of properties of petroleum products: Gasoline (ASTM distillation set-up, specific gravity), Kerosene (flash and fire point, smoke point, aniline point) Diesel (cloud and pour point, aniline point, flash point, viscosity) wax (drop point) Lubricating oil.

About this Chapter

At the end of this chapter, students will be able to:

- Describe specified treatment of the specified type of crude.
- Describe with sketches the procedure to measure the specified property (flash point, fire point, cloud and pour point, drop point, smoke point or aniline point) of the specified petroleum product.
- Explain with sketches the specified type of distillation of crude oil.
- State major properties of the specified petroleum product.
- Explain significance of boiling point range for the specified material.

2.1 INTRODUCTION [S-14]

"A refinery is composed of a group of chemical engineering unit processes and unit operations used for refining certain materials or converting raw material into products of value".

The problem with crude oil is that, it contains hundreds of different types of hydrocarbons all mixed together. You have to separate the different types of hydrocarbons to have anything useful. Fortunately there is an easy way to separate things, and this is what **oil refining** is all about.

Types of Refineries :

The various types of refineries include :

- **Oil refinery:** Converts petroleum crude oil into high-octane motor fuel (gasoline/petrol), diesel oil, liquefied petroleum gases (LPG), jet aircraft fuel, kerosene, heating fuel oils, lubricating oils, asphalt and petroleum coke.

- **Sugar refinery:** Converts sugar cane and sugar beets into crystallized sugar and sugar syrups.
- **Natural gas processing plant:** Purifies and converts raw natural gas into residential, commercial and industrial fuel gas, and also recovers natural gas liquids (NGL) such as ethane, propane, butanes and pentanes.
- **Salt refinery:** Cleans salt (NaCl), produced by the solar evaporation of sea water, followed by washing and re-crystallization.
- Various metal refineries such as alumina, copper, gold, lead, nickel, silver, uranium and zinc.

A typical oil refinery depicts the various **unit** processes and the flow of intermediate product streams that occurs between the inlet crude oil feedstock and the final end products. The diagram depicts only one of the literally hundreds of different oil refinery configurations. It does not include any of the usual refinery facilities providing utilities such as steam, cooling water, and electric power as well as storage tanks for crude oil feedstock and for intermediate products and end products.

2.1.1 The Preliminary Process of Petroleum Refining [S-17]

As mentioned previously, a barrel of crude oil has a mixture of all sorts of hydrocarbons in it. Oil refining separates everything into useful substances.

Chemists use the following steps:

1. The oldest and most common way to separate things into various components (called **fractions**), is to do it using the differences in boiling temperature. This process is called **fractional distillation**. You basically heat crude oil up, let it vapourize and then condense the vapour.
2. Newer techniques use **Chemical processing** on some of the fractions to make others, in a process called **conversion**. Chemical processing, for example, can break longer chains into shorter ones. This allows a refinery to turn diesel fuel into gasoline depending on the demand for gasoline.
3. Refineries must **treat** the fractions to remove impurities.
4. Refineries **combine** the various fractions (processed, unprocessed) into mixtures to make desired products. For example, different mixtures of chains can create gasolines with different **octane ratings**.

The products are stored on-site until they can be delivered to various markets such as gas stations, airports and chemical plants. In addition to making the oil-based products, refineries must also treat the wastes involved in the processes to minimize air and water pollution.

Because crude oil is a mixture of hydrocarbons with different boiling temperatures, it can be separated by distillation into groups of hydrocarbons that boil between two specified boiling points. Two types of distillation are performed: atmospheric and vacuum.

Atmospheric distillation takes place in a distilling column at or near atmospheric pressure. The crude oil is heated to 350 - 400°C and the vapour and liquid are piped into the distilling column. The liquid falls to the bottom and the vapour rises, passing through a series of perforated trays (sieve trays). Heavier hydrocarbons condense more quickly and settle on lower trays and lighter hydrocarbons remain as a vapour longer and condense on higher trays.

Liquid fractions are drawn from the trays and removed. In this way, the light gases, methane, ethane, propane and butane pass out from the top of the column, petrol is formed in the top trays, kerosene and gas oils in the middle and fuel oils at the bottom trays. Residue drawn at the bottom may be burned as fuel, processed into lubricating oils, waxes and bitumen or used as feedstock for cracking units.

To recover additional heavy distillates from this residue, it may be piped to a second distillation column where the process is repeated under vacuum, called **vacuum distillation**. This allows heavy hydrocarbons with boiling points of 450°C and higher to be separated without them partly cracking into unwanted products such as coke and gas.

The heavy distillates recovered by vacuum distillation can be converted into lubricating oils by a variety of processes. The most common of these is called **solvent extraction**. In one version of this process, the heavy distillate is washed with a liquid which does not dissolve in it but which dissolves (and so extracts) the non-lubricating oil components out of it. Another version uses a liquid which does not dissolve in it but which causes the non-lubricating oil components to precipitate (as an extract) from it. Other processes exist which remove impurities by adsorption onto a highly porous solid or which remove any waxes that may be present by causing them to crystallize and precipitate out.

2.1.2 Characterization of the Petroleum

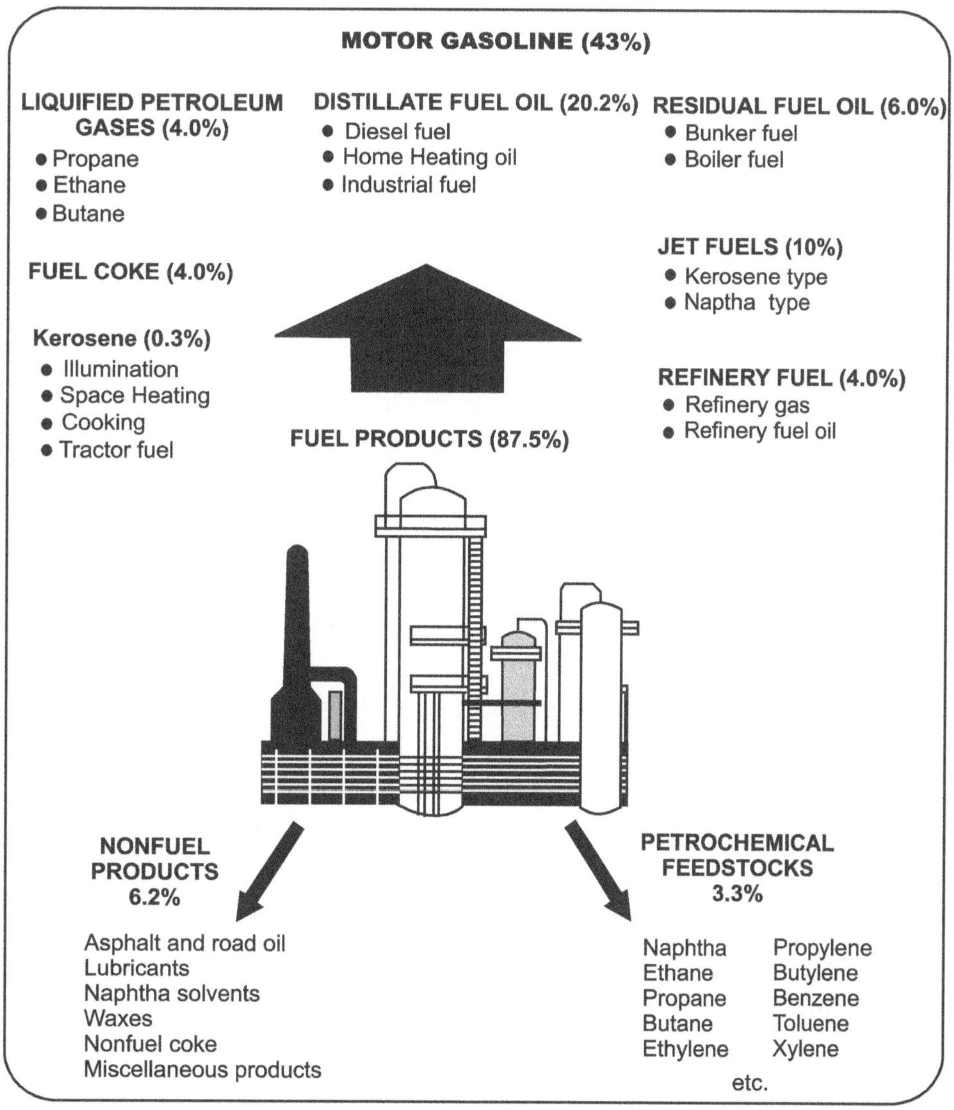

Fig. 2.1 : U.S. refinery products and yields

Product Characterization :

Petroleum refining is the physical, thermal and chemical separation of crude oil into its major distillation fractions which are then further processed through a series of separation and conversion steps into finished petroleum products. The primary products of the industry fall into three major categories: fuels (motor gasoline, diesel and distillate fuel oil, liquefied petroleum gas, jet fuel, residual fuel oil, kerosene, and coke); finished non-fuel products (solvents, lubricating oils, greases, petroleum wax, petroleum jelly, asphalt and coke); and chemical industry feedstocks (naphtha, ethane, propane, butane, ethylene, propylene, butylenes, butadiene, benzene, toluene and xylene). These petroleum products comprise about 40% of the total energy consumed in the U.S. (based on BTUs consumed) and are used as primary input to a vast number of products, including: fertilizers, pesticides, paints, waxes, thinners, solvents, cleaning fluids, detergents, refrigerants, anti-freeze, resins, sealants, insulations, latex, rubber compounds, hard plastics, plastic sheeting, plastic foam and synthetic fibers. About 90% of the petroleum products used in the U.S. are fuels with motor gasoline accounting for about 43% of the total.

2.2 UNIT OPERATIONS USED IN SEPARATION PROCESS

In chemical engineering and related fields, a **unit operation** is a basic step in a process. For example in milk processing, homogenization, pasteurization, chilling, and packaging are each unit operations which are connected to create the overall process. A process may have many unit operations to obtain the desired product.

Historically, the different chemical industries were regarded as different industrial processes and with different principles. In 1923 William H. Walker, Warren K. Lewis and William H. McAdams wrote the book *'The Principles of Chemical Engineering'* and explained the variety of chemical industries used processes which follow the same physical laws. They summed-up these similar processes into unit operations. Each unit operation follows the same physical laws and may be used in all chemical industries. The unit operations form the fundamental principles of chemical engineering.

Chemical engineering **unit operations** consist of five classes:

1. Fluid flow processes, including fluids transportation, filtration, solids, fluidization.
2. Heat transfer processes, including evaporation, condensation.
3. Mass transfer processes, including gas absorption, distillation, extraction, adsorption, drying.
4. Thermodynamic processes, including gas liquefaction, refrigeration.
5. Mechanical processes, including solid transportation, crushing and pulverization, screening and sieving.

Chemical engineering **unit operations** also fall in the following categories:

- Combination (**mixing**)
- Separation (**distillation**)
- Reaction (**chemical reaction**)

Chemical engineering unit operations and chemical engineering unit processing form the main principles of all kinds of chemical industries and are the foundation of designs of chemical plants, factories and equipment used.

The Equipment Used in Refineries :

Refineries utilize a great many different types of physical equipment such as:

- Centrifuges
- Compressors

- Cooling towers
- Crushers
- Crystallizers
- Distillation towers and other pressure vessels
- Electric power generators, transformers and electric motors
- Electrolysis cells
- Evaporators
- Filters
- Furnaces
- Gas flares
- Mixers and blenders
- Monitoring and control systems
- Piping and valves
- Pumps
- Steam generators
- Steam turbines and gas turbines
- Storage tanks
- Wastewater treatment

2.3 PRIMARY TREATMENT OF CRUDE

Dehydration and Desalting of Crude Oil: **[S-17, 18; W-17]**

Crude oil contain a small amount of moisture and salts, that needs to be reduced to a concentration around 5-10 PTB (Pounds per Thousand Barrels) of salt. This salt needs to be removed so that the crude oil can be processed in a refinery or stabilizer plant without fouling heaters, exchangers and other equipment. Water is likely to occur in emulsion form when the crudes are naphthenic or sulphurous. No harm may be expected to the distillation column due to the presence of moisture, as there is always steam in distillation. However, crude has to be dehydrated to remove the salts. Water being good solvent for these salts, the removal is very much effective in the form of brine. Of all the existing salts, chlorides of calcium and magnesium distinguish themselves in playing an invincible role in overhead corrosion. These salts in presence of steam at 150-200°C easily hydrolise generating hydrochloric vapours. These vapours cause corrosion to the equipment. Any crude that contains more than 5 kgs of total salts expressed in terms of sodium chloride per thousand barrels may be regarded as salty crude.

In general, dehydration removes all salts which is practiced out in two stages; first at the site of mine and later in the refinery.

The following general methods are versatile for dehydration of crudes.

1. Chemical treatment
2. Gravity settling
3. Centrifugal separation
4. Electric desalter

At the oil field, salt is removed by settling or by adding chemicals or by combination of these two. Crudes possessing emulsifying characteristics are not responsive to the settling method; these demand demulsifying agents to increase the coalescence of water drops. Soda ash, sodium hydroxide, salts of fatty acids, petroleum sulfonates are such chemicals; which hasten the agglomeration of water droplets. A good amount of water should be available in crude for such treatment; lack of water demands fresh additions to the extent of 20% even. After adding the chemicals and water to desire extents the crude mixture is allowed to stand at 75-80°C, at a pressure of 15 kgs/cm^2 in huge tall tanks. Pressure ensure the retention of volatiles in crude. Demulsifying chemicals, if necessary are added in very small amount. Good separation into hydrocarbon and aqueous phase results when the crude is allowed to stand for 48 hours.

Coalescence is also aided by passing the mixture through towers packed with gravel. The settling techniques are not effective and time consuming. Continuous operations are not possible, with the result large amount of space and equipment are to be isolated for this purpose. Similarly centrifugal separation is also not economical due to huge energy requirements and less quantity handled; all these have given the way to electric desalting.

2.3.1 Electric Desalting [S-16]

Before separation into fractions, crude oil usually must first be treated to remove corrosive salts. The desalting process also removes some of the metals and suspended solids which cause catalyst deactivation. Desalting involves the mixing of heated crude oil with water (about 3% to 10% of the crude oil volume) so that the salts are dissolved in the water. The water must then be separated from the crude oil in a separating vessel by adding demulsifier chemicals to assist in breaking the emulsion and/or, more commonly, by applying a high potential electric field across the settling vessel to coalesce the polar salt water droplets. The desalting process creates an oily desalter sludge and a high temperature salt water waste stream which is typically added to other process wastewaters for treatment in the refinery wastewater treatment facilities. The water used in crude desalting is often untreated or partially treated water from other refining process water sources.

CRUDE OIL PRETREATMENT

Description :

(a) Crude oil often contains water, inorganic salts, suspended solids and water-soluble trace metals. As a first step in the refining process, to reduce corrosion, plugging, and fouling of equipment, and to prevent poisoning the catalysts in processing units, these contaminants must be removed by desalting (dehydration).

(b) The two most typical methods of crude-oil desalting, chemical and electrostatic separation, use hot water as the extraction agent. In chemical desalting, water and chemical surfactants (demulsifiers) are added to the crude, heated so that salts and other impurities dissolve into the water or attach to the water, and then held in a tank where they settle out. Electrical desalting is the application of high-voltage electrostatic charges to concentrate suspended water globules in the bottom of the settling tank. Surfactants are added only when the crude has a large amount of suspended solids. Both methods of desalting are continuous. A third and less-common process involves filtering heated crude using diatomaceous earth.

(c) The feedstock crude oil is heated between 66°C and 177°C to reduce viscosity and surface tension for easier mixing and separation of the water. The temperature is limited by the vapour pressure of the crude-oil feedstock. In both methods, other chemicals may be added. Ammonia is often used to reduce corrosion. Caustic or acid may be added to adjust the pH of the water wash. Wastewater and contaminants are discharged from the bottom of the settling tank to the wastewater treatment facility. The desalted crude is continuously drawn from the top of the settling tanks and sent to the crude distillation (fractionating) tower.

Table 2.1 : Desalting process

Feedstock	From	Process	Typical Products	To
Crude	Storage	Treating	Desalted crude	Atmospheric distillation tower
			Wastewater	Treatment

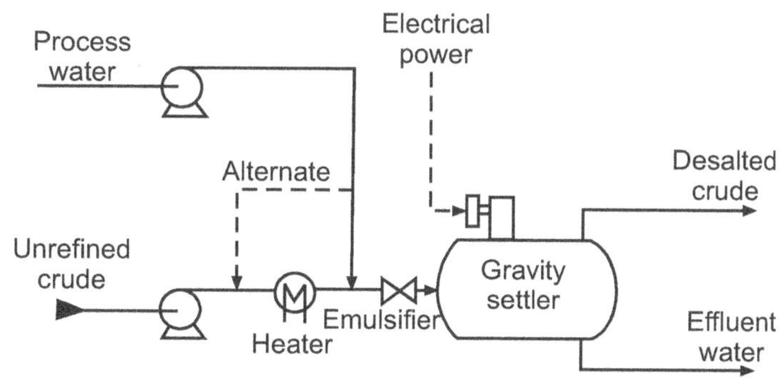

Fig. 2.2 : Electric desalting

2.4 DISTILLATION OF CRUDE [S-14]

Introduction : [S-16, 17]

The first step in the refining process is the separation of crude oil into various fractions or straight-run cuts by distillation in atmospheric and vacuum towers. The main fractions or "cuts" obtained have specific boiling-point ranges and can be classified in order of decreasing volatility into gases, light distillates, middle distillates, gas oils and residue.

The various components of crude oil have different sizes, weights and boiling temperatures; so, the first step is to separate these components. Because they have different boiling temperatures, they can be separated easily by a process called **fractional distillation**.

The steps of fractional distillation are as follows :

1. You **heat** the mixture of two or more substances (liquids) with different boiling points to a high temperature. Heating is usually done with high pressure steam to temperatures of about 600° Celsius.
2. The mixture **boils**, forming vapour (gases); most substances go into the vapour phase.
3. The **vapour** enters the bottom of a long column (**fractional distillation column**) that is filled with trays or plates.
   (a) The trays have many holes or bubble caps (like a loosened cap on a soda bottle) in them to allow the vapour to pass through.
   (b) The trays increase the contact time between the vapour and the liquids in the column.
   (c) The trays help to collect liquids that form at various heights in the column.
   (d) There is a temperature difference across the column (hot at the bottom, cool at the top).
4. The **vapour** rises in the column.
5. As the vapour rises through the trays in the column, it **cools**.

6. When a substance in the vapour reaches a height where the temperature of the column is equal to that substance's boiling point, it will **condense** to form a liquid. (The substance with the lowest boiling point will condense at the highest point in the column; substances with higher boiling points will condense lower in the column.)
7. The trays **collect** the various liquid fractions.
8. The collected liquid fractions may:
   - pass to condensers, which cool them further, and then go to storage tanks.
   - go to other areas for further chemical processing.

Fractional distillation is useful for separating a mixture of substances with narrow differences in boiling points, and is the most important step in the refining process.

(I) Atmospheric Distillation of Crude: [S-14, 16, 18; W-13, 17]

(a) At the refinery, the desalted crude feedstock is preheated using recovered process heat. The feedstock then flows to a direct-fired crude charge heater where it is fed into the vertical distillation column just above the bottom, at pressures slightly above atmospheric pressure and at temperatures ranging from 340°C to 370°C (heating crude oil above these temperatures may cause undesirable thermal cracking). All but the heaviest fractions flash into vapour. As the hot vapour rises in the tower, its temperature is reduced. Heavy fuel oil or asphalt residue is taken from the bottom. At successively higher points on the tower, the various major products including lubricating oil, heating oil, kerosene, gasoline and uncondensed gases (which condense at lower temperatures) are drawn off.

(b) The fractionating tower, a steel cylinder about 120 feet high, contains horizontal steel trays for separating and collecting the liquids. At each tray, vapours from below enter perforations and bubble caps. They permit the vapours to bubble through the liquid on the tray, causing some condensation at the temperature of that tray. An overflow pipe drains the condensed liquids from each tray back to the tray below, where the higher temperature causes re-evaporation. The evaporation, condensing, and scrubbing operation is repeated many times until the desired degree of product purity is reached. Then side streams from certain trays are taken off to obtain the desired fractions. Products ranging from uncondensed fixed gases at the top to heavy fuel oils at the bottom can be taken continuously from a fractionating tower. Steam is often used in towers to lower the vapour pressure and create a partial vacuum. The distillation process separates the major constituents of crude oil into so-called straight-run products. Sometimes crude oil is "topped" by distilling off only the lighter fractions, leaving a heavy residue that is often distilled further under high vacuum.

Table 2.2 : Atmospheric distillation process

Feedstock	From	Process	Typical Products	To
Crude	Desalting	Separation	Gases	Atmospheric distillation tower
			Napthas	Reforming or treating
			Kerosene or distillates	Treating
			Gas oil	Catalytic cracking
			Residual	Vacuum tower or visbreaker

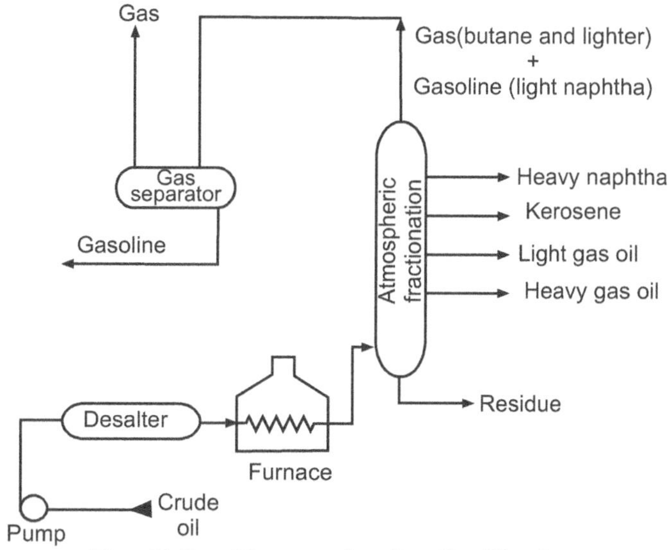

Fig. 2.3 : Atmospheric distillation

(II) Vacuum Distillation of Crude: [S-16, 17, 18; W-13, 17]

In order to further distill the residuum or topped crude from the atmospheric tower at higher temperatures, reduced pressure is required to prevent thermal cracking. The process takes place in one or more vacuum distillation towers.

"The principles of vacuum distillation resemble those of fractional distillation and, except that larger-diameter columns are used to maintain comparable vapour velocities at the reduced pressures", the equipment is also similar.

To recover additional distillates from long residue, distillation at reduced pressure and high temperature has to be applied. This vacuum distillation process has become an important chain in maximising the upgrading of crude oil. As distillates, vacuum gas oil, lubricating oils and/or conversion feedstocks are generally produced. The residue from vacuum distillation - short residue - can be used as feedstock for further upgrading, as bitumen feedstock or as fuel component. The technology of vacuum distillation has developed considerably in recent decades. The main objectives have been to maximize the recovery of valuable distillates and to reduce the energy consumption of the units.

Table 2.3 : Vacuum distillation process

Feedstock	From	Process	Typical Products	To
Residuals	Atmospheric tower	Separation	Gas oils	Catalytic cracker
			Lubricants	Hydrotreating or solvent
			Residual	Deasphalter, visbreaker, or coker

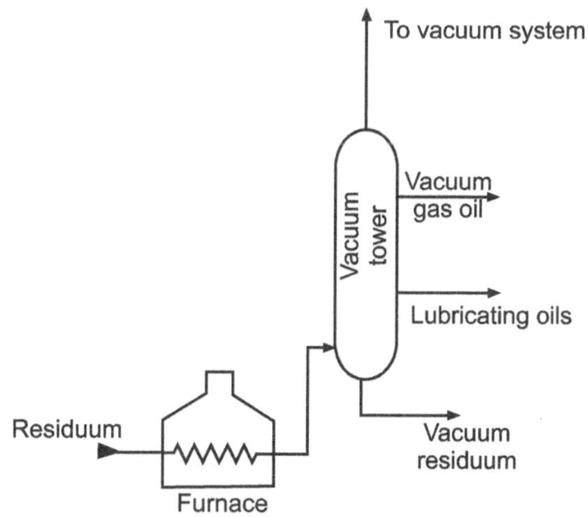

Fig. 2.4 : Vacuum distillation

Heavier fractions from the atmospheric distillation unit that cannot be distilled without cracking under its pressure and temperature conditions are vacuum distilled. Vacuum distillation is simply the distillation of petroleum fractions at a very low pressure (0.2 to 0.7 psia) to increase volatilization and separation. In most systems, the vacuum inside the fractionator is maintained with steam ejectors and vacuum pumps, barometric condensers or surface condensers. The injection of superheated steam at the base of the vacuum fractionator column further reduces the partial pressure of the hydrocarbons in the tower, facilitating vapourization and separation. The heavier fractions from the vacuum distillation column are processed downstream into more valuable products through either cracking or coking operations.

A potential source of emissions from distillation of crude oil are the combustion of fuels in the furnace and some light gases leaving the top of the condensers on the vacuum distillation column. A certain amount of non-condensable light hydrocarbons and hydrogen sulfide pass through the condenser to a hot well, and then are discharged to the refinery sour fuel system or are vented to a process heater, flare or another control device to destroy hydrogen sulfide. The quantity of these emissions depends on the size of the unit, the type of feedstock, and the cooling water temperature. If barometric condensers are used in vacuum distillation, significant amounts of oily wastewater can be generated. Vacuum pumps and surface condensers have largely replaced barometric condensers in many refineries to eliminate this oily wastewater stream. Oily sour water is also generated in the fractionators.

The internal designs of some vacuum towers are different from atmospheric towers, in that, random packing and demister pads are used instead of trays. A typical first-phase vacuum tower may produce gas oils, lubricating-oil base stocks, and heavy residual for propane deasphalting. A second-phase tower operating at lower vacuum may distill surplus residuum from the atmospheric tower, which is not used for lube-stock processing, and surplus residuum from the first vacuum tower not used for deasphalting. Vacuum towers are typically used to separate catalytic cracking feedstock from surplus residuum.

At the place where the heated feed is introduced in the vacuum column - called the flash zone - the temperature should be high and the pressure as low as possible to obtain maximum distillate yield. The flash temperature is restricted to about 420°C, however, in view of the cracking tendency of high-molecular-weight hydrocarbons. Vacuum is maintained with vacuum ejectors and lately also with liquid ring pumps. Lowest achievable vacuum in the flash zone is in the order of 10 mbar.

In the older type high vacuum units the required low hydrocarbon partial pressure in the flash zone could not be achieved without the use of "lifting" steam. The steam acts in a similar manner as the stripping steam of crude distillation units. This type of units is called **"wet units"**. One of the latest developments in vacuum distillation has been the deep vacuum flashers, in which no steam is required. These **"dry units"** operate at very low flash zone pressures and low pressure drops over the column internals. For that reason the conventional reflux sections with fractionation trays have been replaced by low pressure-drop spray sections. Cooled reflux is sprayed via a number of specially designed spray nozzles in the column countercurrent to the up-flowing vapour. This spray of small droplets comes into close contact with the hot vapour, resulting in good heat and mass transfer between the liquid and vapour phase. To achieve low energy consumption, heat from the circulating refluxes and rundown streams is used to heat up the long residue feed. Surplus heat is used to produce medium and/or low-pressure steam or is exported to another process unit (via heat integration). The direct fuel consumption of a modern high-vacuum unit is approximately 1% on intake, depending on the quality of the feed. The steam consumption of the dry high-vacuum units is significantly lower than that of the "wet" units. They have become net producers of steam instead of steam consumers.

(III) Other Distillation Towers (Columns) :

Within refineries there are numerous other, smaller distillation towers called columns, designed to separate specific and unique products. Columns all work on the same principles as the towers described above. For example, a depropanizer is a small column designed to separate propane and lighter gases from butane and heavier components. Another larger column is used to separate ethyl benzene and xylene. Small "bubble" towers called strippers use steam to remove trace amounts of light products from heavier product streams.

(IV) Solvent Extraction :

Introduction :

Solvent treating is a widely used method of refining lubricating oils as well as a host of other refinery stocks. Since distillation (fractionation) separates petroleum products into groups only by their boiling-point ranges, impurities may remain. These include organic compounds containing sulfur, nitrogen and oxygen; inorganic salts and dissolved metals; and soluble salts that were present in the crude feedstock. In addition, kerosene and distillates may have trace amounts of aromatics and naphthenes, and lubricating oil base-stocks may contain wax. Solvent refining processes including solvent extraction and solvent dewaxing usually remove these undesirables at intermediate refining stages or just before sending the product to storage.

Explanation :

(a) The purpose of solvent extraction is to prevent corrosion, protect catalyst in subsequent processes, and improve finished products by removing unsaturated, aromatic hydrocarbons from lubricant and grease stocks. The solvent extraction process separates aromatics, naphthenes, and impurities from the product stream by dissolving or precipitation. The feedstock is first dried and then treated using a continuous countercurrent solvent treatment operation. In one type of process, the feedstock is washed with a liquid in which the substances to be removed are more soluble than in the desired resultant product. In another process, selected solvents are added to cause impurities to precipitate out of the product. In the adsorption process, highly porous solid materials collect liquid molecules on their surfaces.

(b) The solvent is separated from the product stream by heating, evaporation, or fractionation, and residual trace amounts are subsequently removed from the raffinate by steam stripping or vacuum flashing. Electric precipitation may be used for separation of inorganic compounds. The solvent is then regenerated to be used again in the process.

(c) The most widely used extraction solvents are phenol, furfural, and cresylic acid. Other solvents less frequently used are liquid sulfur dioxide, nitrobenzene, and 2,2'-dichloroethyl ether. The selection of specific processes and chemical agents depends on the nature of the feedstock being treated, the contaminants present, and the finished product requirements.

Table 2.4 : Solvent extraction process

Feedstock	From	Process	Typical Products	To
Naphthas, distillates, kerosene	Atmospheric tower	Treating/ blending	High octane gasoline	Storage
			Refined fuels	Treating and blending
			Spent agents	Treating and blending

(v) Absorption Process :

- This process selectively removes a certain gas from a gas mixture using a liquid absorbent.
- In the refining industry, this process is used extensively to free the product gas streams from acid gases (mainly H_2S) either by using a physical or a chemical absorbent.

(VI) Adsorption Process :

- Adsorption processes use a solid material (adsorbent) possessing a large surface area and the ability to selectively adsorb a gas or a liquid on its surface.
- Example : Silica (SiO_2), anhydrous alumina (Al_2O_3), and molecular sieves (crystalline silica/alumina).
- Adsorption of n-paraffins ($C_{10}-C_{14}$) from a kerosine or a gas oil fraction can be achieved in a liquid or a vapour phase adsorption process.
- The Iso-Siv process is an isobaric, isothermal adsorption technique used to separate n-paraffins from gas oils. The operation conditions are approximately 370°C and 100 psi. Desorption is achieved using n-pentane or n-hexane. The solvent is easily distilled from the heavier n-paraffins and then recycled.

2.5 HYDROCARBONS/FRACTIONS OBTAINED FROM CRUDE OIL, BOILING RANGE AND THEIR USES

Every refinery begins with the separation of crude oil into different hydrocarbons or fractions by distillation. [S-14, 16, 17, 18; W-17]

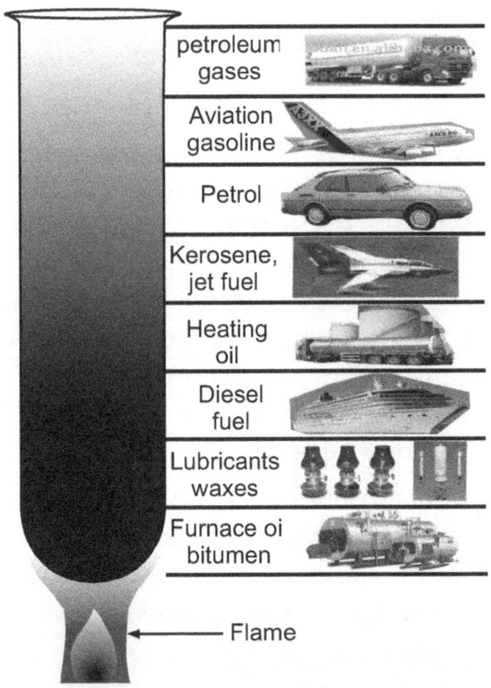

Fig. 2.5 : Refining product

The fractions are further treated to convert them into mixtures of more useful saleable products by various methods such as cracking, reforming, alkylation, polymerization and isomerization. These mixtures of new compounds are then separated using methods such as fractionation and solvent extraction. Impurities are removed by various methods, e.g. dehydration, desalting, sulphur removal and hydrotreating.

Refinery processes have developed in response to changing market demands for certain products. With the advent of the internal combustion engine, the main task of refineries became the production of petrol.

The quantities of petrol available from distillation alone was insufficient to satisfy consumer demand. Refineries began to look for ways to produce more and better quality petrol.

Two types of processes have been developed :

- Breaking down large, heavy hydrocarbon molecules.
- Reshaping or rebuilding hydrocarbon molecules.

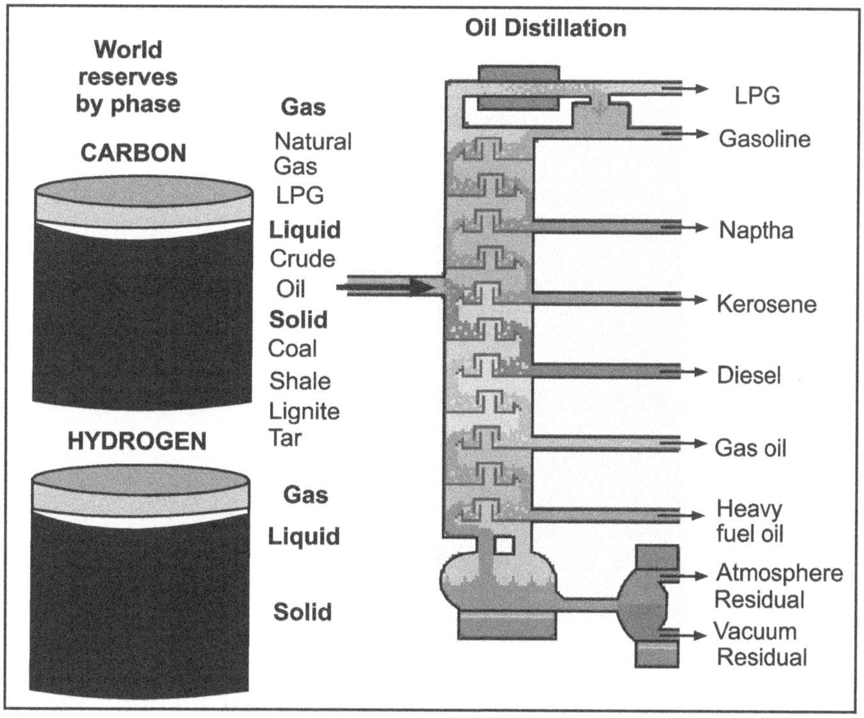

Fig. 2.6 : Crude oil refining

Table 2.5 : Distillate fraction with their boiling point and carbon atoms per molecule

Crude Oil Refining		
Distillate Fraction	**Boiling Point (°C)**	**Carbon Atoms per Molecule**
Gases	Below 30	1–4
Gasoline	30–210	5–12
Naphtha	100–200	8–12
Kerosene and Jet Fuel	150–250	11–13
Diesel and Fuel Oil	160–400	13–17
Atmospheric Gas Oil	220–345	–
Heavy Fuel Oil	315–540	20–45
Atmospheric Residue	Over 450	Over 30
Vacuum Residue	Over 615	Over 60

Petroleum products comprise refinery gas, ethane, LPG, aviation gasoline, motor gasoline, jet fuels, kerosene, gas/diesel oil, heavy fuel oil, naphtha, white spirit, lubricants, bitumen, paraffin waxes, petroleum coke and other petroleum products. Petroleum products are any oil-based products which can be obtained by distillation and are normally used outside the refining industry. The exception to this are those finished products which are classified as refinery feedstocks above. Production of the petroleum products shows gross refinery output for each product. Refinery fuel (row petroleum refineries, under energy sector) represents consumption of petroleum products, both intermediate and finished, within refineries, e.g. for heating, lighting, traction etc.

The different products and uses of petroleum product : **(W-13)**

- Aviation Gasoline
- Gas/Diesel Oil (Distilled Fuel Oil)

- Heavy Fuel Oil Residual
- Kerosene
- Jet Fuel
- Liquefied Petroleum Gases (LPG) and Ethane
- Motor Gasoline
- Naphtha
- Other Petroleum Products
- Petroleum Coke
- Refinery Gas (not liquefied)
- A.T.F. (Aviation Turbine Fuel)
- HSD (High Speed Diesel)
- LDO (Light Diesel Oil)
- FO (Fuel Oil)
- Bitumen

1. Aviation Gasoline :

Aviation gasoline is motor spirit prepared especially for aviation piston engines, with an octane number suited to the engine, a freezing point of −60°C, and a distillation range usually within the limits of 30°C and 180°C.

2. Gas Diesel Oil/(Distillate Fuel Oil) :

Gas/diesel oil includes heavy gas oils. Gas oils are obtained from the lowest fraction from atmospheric distillation of crude oil, while heavy gas oils are obtained by vacuum redistillation of the residual from atmospheric distillation. Gas/diesel oil distills between 180°C and 380°C. Several grades are available depending on uses: diesel oil for diesel compression ignition (cars, trucks, marine etc.), light heating oil for industrial and commercial uses, and other gas oil including heavy gas oils which distil between 380°C and 540°C and which are used as petrochemical feedstocks.

3. Heavy Fuel Oil Residual :

This heading defines oils that make up the distillation residue. It comprises all residual fuel oils (including those obtained by blending). Its kinematic viscosity is above 10 cST at 80°C. The flash point is always above 50°C and the density is always more than 900 kg/l.

4. Kerosene :

Kerosene comprises refined petroleum distillate intermediate in volatility between gasoline and gas/diesel oil. It is a medium oil, distilling between 150°C and 300°C.

While in India, kerosene is largely used for illumination elsewhere it is used for heating purposes. A special cut of kerosene can be processed for the manufacture of linear alkyl benzene used as a detergent.

5. Jet Fuel :

This category comprises both gasoline and kerosene type jet fuels meeting specifications for use in aviation turbine power units.

(a) Gasoline type jet fuel : This includes all light hydrocarbon oils for use in aviation turbine power units. They distill between 100°C and 250°C. It is obtained by blending kerosenes and gasoline or naphthas in such a way that the aromatic content does not exceed 25% in volume, and the vapour pressure is between 13.7 kPa and 20.6 kPa. Additives can be included to improve fuel stability and combustibility.

(b) Kerosene type jet fuel : This is medium distillate used for aviation turbine power units. It has the same distillation characteristics and flash point as kerosene (between 150°C and 300°C but not generally above 250°C). In addition, it has particular specifications (such as freezing point) which are established by the International Air Transport Association (IATA).

6. LPG (Liquified Petroleum Gas) :

These are the light hydrocarbons fraction of the paraffin series, derived from refinery processes, crude oil stabilization plants and natural gas processing plants comprising propane (C_3H_8) and butane (C_4H_{10}) or a combination of the two. They are normally liquefied under pressure for transportation and storage. Ethane is a naturally gaseous straight-chain hydrocarbon (C_2H_6). It is a colourless paraffinic gas which is extracted from natural gas and refinery gas streams.

This is generally a mixture of propane and butane. At atmospheric temperature and pressure it is gas. It has therefore to be stored under pressure to keep it liquified; hence the name LPG.

LPG is used largely as a domestic fuel. It is also invaluable for use as a clean and convenient fuel in certain industries, such as glass manufacture. LPG is also a good feedstock for the manufacture of various petrochemicals.

7. Motor Gasoline :

This is light hydrocarbon oil for use in internal combustion engines such as motor vehicles, excluding aircraft. Motor gasoline is distilled between 35°C and 215°C and is used as a fuel for land based spark ignition engines. Motor gasoline may include additives (such as ethanol), oxygenates and octane enhancers, including lead compounds such as TEL (Tetraethyl lead) and TML (Tetramethyl Lead).

This is a blend of a variety of light hydrocarbons such as straight run naphtha, reformate, cracked gasoline from crackers, alkylates etc. Octane rating of motor gasoline is an important anti-knock property which makes the modern high compression ratio of internal combustion engines possible for automotive purpose.

8. Naphtha :

Naphtha is a feedstock destined for the petrochemical industry (e.g. ethylene manufacture or aromatics production). Naphtha comprises material in the 30°C and 210°C distillation range or part of this range.

The most important usage of naphtha is in the production of Motor Spirit. It is partly blended direct into M.S. and partly processed in a catalytic reforming unit to produce high octane number reformate. Reformate is then blended into the M.S. pool. Naphtha is also an important feedstock for fertilizer manufacture.

Steam cracking of naphtha produces a large variety of olefins including ethylene and propylene which are the precursors for various petrochemicals.

9. Other Petroleum Products :

White Spirit and SBP : White spirit and SBP are refined distillate intermediates with a distillation in the naphtha/kerosene range. They are sub-divided as :

(a) Industrial Spirit (SBP) : Light oils distilling between 30°C and 200°C, with a temperature difference between 5% volume and 90% volume distillation points, including losses, of not more than 60°C. In other words, SBP is a light oil of narrower cut than motor spirit. There are 7 or 8 grades of industrial spirit, depending on the position of the cut in the distillation range defined above.

(b) **White Spirit :** Industrial spirit with a flash point above 30°C. The distillation range of white spirit is 135°C to 200°C.

(c) **Lubricants :** Lubricants are hydrocarbons produced from distillate or residue; they are mainly used to reduce friction between bearing surfaces. This category includes all finished grades of lubricating oil, from spindle oil to cylinder oil, and those used in greases, including motor oils and all grades of lubricating oil base stocks.

(d) **Bitumen :** Solid, semi-solid or viscous hydrocarbon with a colloidal structure, being brown to black in colour, obtained as a residue in the distillation of crude oil, vacuum distillation of oil residues from atmospheric distillation. Bitumen is often referred to as asphalt and is primarily used for surfacing of roads and for roofing material. This category includes fluidised and cut back bitumen.

(e) **Paraffin Waxes :** Saturated aliphatic hydrocarbons (with the general formula C_nH_{2n+2}). These waxes are residues extracted when dewaxing lubricant oils and they have a crystalline structure with carbon number greater than 12. Their main characteristics are as follows: they are colourless, odourless and translucent, with a melting point above 45°C.

(f) **Others :** Includes the petroleum products not classified above, for example: tar, sulphur and grease. This category also includes aromatics (e.g. BTX or benzene, toluene and xylene) and olefins (e.g. propylene) produced within refineries.

10. Petroleum Coke :

Petroleum coke is defined as 'a black solid residue, obtained mainly by cracking and carbonising of residue feedstocks, tar and pitches in processes such as delayed coking or fluid coking'. It consists mainly of carbon (90 to 95%) and has a low ash content. It is used as a feedstock in coke ovens for the steel industry, for heating purposes, for electrode manufacture and for production of chemicals. The two most important qualities are "green coke" and "calcinated coke". This category also includes "catalyst coke" deposited on the catalyst during refining processes. This coke is not recoverable and is usually burned as refinery fuel.

11. Refinery Gas :

Refinery gas is defined as 'a non-condensible gas obtained during distillation of crude oil or treatment of oil products (e.g. cracking) in refineries'. It consists mainly of hydrogen, methane, ethane and olefins. It also includes gases which are returned from the petrochemical industry. Refinery gas production refers to gross production. Own consumption is shown separately under petroleum refineries in the energy sector.

12. A.T.F. (Aviation Turbine Fuel) :

Aviation turbine fuel also called the jet fuel is becoming an increasingly important fuel with the development of aviation industry and with the phasing out of the earlier piston type of aero-engines. This is a kind of a kerosene cut with certain important specifications such as the freezing point of ATF should be bellow 47°C.

13. HSD (High Speed Diesel) :

This is an important petroleum product used for automotive purposes in compression ignition engines. It is an important fuel used for road, rail and marine transport systems.

This is an inferior kind of a gas oil used as fuel in low speed engines largely used in lift irrigation sector in India.

14. FO (Fuel Oil) :

This is used in a variety of industrial applications as fuel in furnaces.

15. Bitumen :

The bulk grades of bitumen are used for road making.

2.6 PROPERTIES OF PETROLEUM PRODUCTS

Crude oils differ appreciably in their properties according to origin and the ratio of the different components in the mixture. Lighter crudes generally yield more valuable light and middle distillates and are sold at higher prices. Crudes containing a high % of impurities, such as sulfur compounds, are less desirable than low-sulfur crudes because of their corrosivity and the extra treating cost. Corrosivity of crude oils is a function of many parameters among which are the type of sulfur compounds and their decomposition temperatures, the total acid number, the type of carboxylic and naphthenic acids in the crude and their decomposition temperatures. It was found that naphthenic acids begin to decompose at 600°F. Refinery experience has shown that above 750°F there is no naphthenic acid corrosion. The subject has been reviewed by Kane and Cayard. For a refiner, it is necessary to establish certain criteria to relate one crude to another to be able to assess crude quality and choose the best processing scheme.

The main important properties of crude oil are: **[S-16, 17, 18; W-17]**

1. **Flash point:** It is the lowest temperature at which the oil gives off enough vapours which ignite for moment, when small flame is brought to near it.
2. **Fire point:** It is the lowest temperature at which vapours of oil burn continuously for at least five seconds, when small flame is brought to near it. Usually fire points are 5°C to 40°C higher than that of flash point.
3. **Volatility:** When oil is used at higher temperature in heavy machineries, a portion of oil may get vapourized. The residual oil left behind may possess quite different properties like decrease in viscosity.
4. **Viscosity:** It is the measure of internal resistance of fluid i.e. its own moving particles. Viscosity is defined as, "force in dynes, required to move one sq.cm. of fluid over a layer of same area and through 1 cm distance with a velocity of 1 cm/sec."
5. **Viscosity Index:** Viscosity of oil decreases with increase in temperature. The rate at which viscosity of oil changes with temperature is measured by arbitrary scale known as Viscosity Index.
6. **Oiliness:** It is the property of oil by virtue of which they stick on the surface of machine part and remain in place even at condition of high speed and heavy load.
7. **Cloud point:** When oil is cooled slowly, the temperature at which it becomes cloudy is called as Cloud point.
8. **Pour point:** The temperature at which oil stops flowing or getting poured is called as Pour point of oil.
9. **Drop point:** In general, the dropping point is the temperature at which the grease passes from a semisolid to a liquid state under the conditions of test. This change in state is typical of greases containing as thickeners soaps of conventional types. Greases containing as thickeners, materials other than conventional soaps can, without change in state, separate oil. This test method is useful to assist in identifying the grease as to type and for establishing and maintaining bench marks for quality control.

10. **Aniline point:** The Aniline point of an oil is defined as 'the minimum temperature at which equal volumes of Aniline ($C_6H_5NH_2$) and the oil are miscible, i.e. form a single phase upon mixing'.

    The value gives an approximation for the content of aromatic compounds in the oil, since the miscibility of aniline, which is also an aromatic compound suggests the presence of similar (i.e. aromatic) compounds in the oil. The lower the aqniline point, the greater is the content of aromatic compounds in the oil as obviously a lower temperature is needed to ensure miscibility.

11. **Cetane number:** Cetane number or CN is a measure of a fuel's ignition delay, the time period between the start of injection and the first identifiable pressure increase during combustion of the fuel. In a particular diesel engine, higher cetane fuels will have shorter ignition delay periods than lower cetane fuels. Cetane numbers are only used for the relatively light distillate diesel oils.

    In short, the higher the cetane number, the more easily the fuel will combust in a compression setting (such as a diesel engine). The characteristic diesel "knock" occurs when the first portion of fuel that has been injected into the cylinder suddenly ignites after an initial delay. Minimizing this delay results in less unburned fuel in the cylinder at the beginning and less intense knock. Therefore higher-cetane fuel usually causes an engine to run more smoothly and quietly. This does not necessarily translate into greater efficiency, although it may be in certain engines.

12. **Octane number:** Octane number is a standard measure of the performance of a motor or aviation fuel. The higher the octane number, the more compression the fuel can withstand before detonating (igniting).

    Octane number, also called Antiknock Rating, measure of the ability of a fuel to resist knocking when ignited in a mixture with air in the cylinder of an internal-combustion engine. The octane number is determined by comparing, under standard conditions, the knock intensity of the fuel with that of blends of two reference fuels: iso-octane, which resists knocking, and heptane, which knocks readily. The octane number is the percentage by volume of iso-octane in the iso-octane–heptane mixture that matches the fuel being tested in a standard test engine.

13. **Ignition temperature:** The ignition temperature is 'the lowest temperature at which a flammable substance in an air-liquid mixture ignites spontaneously and continues to burn without any addition of heat'. Or **[S-16]**

    The ignition temperature is 'the lowest temperature at which the vapour ignites spontaneously from the energy of the environment'.

14. **Calorific value:** It can be defined as, 'the amount of heat liberated in kJ or kcal by the complete combustion of 1 kg of fuel'.

15. **Carbon residue:** Carbon residue is defined as, 'the carbonaceous residue formed after thermal destruction of a sample'.

2.6.1 Test to Determine Properties of Crude Oil

1. **Density:** Density is defined as "the mass of unit volume of a material at a specific temperature". A more useful unit used by the petroleum industry is specific gravity, which is the ratio of the weight of a given volume of a material to the weight of the same volume of water measured at the same temperature.

2. Specific gravity is used to calculate the mass of crude oils and its products.

Usually, crude oils and their liquid products are first measured on a volume basis, then changed to the corresponding masses using the specific gravity.

3. The API (American Petroleum Institute) gravity is another way to express the relative masses of crude oils. The API gravity could be calculated mathematically using the following equation:

$$^{\circ}\text{API} = \frac{141.5}{\text{Specific gravity } 60.60^{\circ}} - 131.5$$

A low API gravity indicates a heavier crude oil or a petroleum product, while a higher API gravity means a lighter crude oil or product. Specific gravities of crude oils roughly range from 0.82 for lighter crudes to over for heavier crudes (41 to 10 °API scale).

4. Salt Content: The salt content expressed in milligrams of sodium chloride per litre oil (or in pounds/barrel) indicates the amount of salt dissolved in water. Water in crudes is mainly present in an emulsified form. A high salt content in a crude oil presents serious corrosion problems during the refining process. In addition, high salt content is a major cause of plugging heat exchangers and heater pipes. A salt content higher than 10 lb/1,000 barrels (expressed as NaCl) requires desalting.

5. Sulfur Content: Determining the sulfur content in crudes is important because the amount of sulfur indicates the type of treatment required for the distillates. To determine sulfur content, a weighed crude sample (or fraction) is burned in an air stream. All sulfur compounds are oxidized to sulfur dioxide, which is further oxidized to sulfur trioxide and finally titrated with a standard alkali.

Identifying sulfur compounds in crude oils and their products is of little use to a refiner because all sulfur compounds can easily be hydrode sulfurized to hydrogen sulfide and the corresponding hydrocarbon. The sulfur content of crudes, however, is important and is usually considered when determining commercial values.

6. Pour Point: The pour point of a crude oil or product is the lowest temperature at which an oil is observed to flow under the conditions of the test. Pour point data indicates the amount of long-chain paraffins (petroleum wax) found in a crude oil. Paraffinic crudes usually have higher wax content than other crude types. Handling and transporting crude oils and heavy fuels is difficult at temperatures below their pour points. Often, chemical additives known as pour point depressants are used to improve the flow properties of the fuel. Long-chain n-paraffins ranging from 16 to 60 carbon atoms in particular, are responsible for near-ambient temperature precipitation.

In middle distillates, less than 1% wax can be sufficient to cause solidification of the fuel.

5. Ash Content: This test indicates the amount of metallic constituents in a crude oil. The ash left after completely burning an oil sample usually consists of stable metallic salts, metal oxides and silicon oxide. The ash could be further analyzed for individual elements using spectroscopic techniques.

2.7 TESTING APPARATUS

2.7.1 Pensky Martens Apparatus

The flash point of any oil is defined as, "the lowest temperature at which it forms vapors and produces combustible mixture with air". The higher flash point temperature is always desirable for any lubricating oil and lower flash point for any fuel. If the oil has the lower value of the flash point temperature, it will burn easily and forms the carbon deposits on the moving parts. The minimum flash temperature of the oil used in I.C engines varies from 200°C to 250°C. When the oil is tested by using the open cup apparatus, the temperature is slightly more than the above temperatures.

The fire point of any oil is the lowest temperature at which oil just begins to burn continuously. Flash and fire point temperature differs by 3°C to 5°C when it is tested. However, a greater difference may be obtained if some additives are mixed with oil. The flash and the fire point temperatures depends upon the volatility of the oil.

Fire and flash points are good indication of relative flammability of the oil and also limits the operating temperature of oil. Except for the safety from fire hazards, they do not have any significance for engine operation. However, fire and flash points of lube oil are very good indication of the crankcase dilution. The fuel, which leak into the crankcase, readily evaporate and burn at considerably lower temperature than the temperature at which the oil would have burned, clearly indicates the degree of dilution.

Flash and fire point of the given light oil (Kerosene/Diesel) is determined by heating the oil in Pensky Martens Closed Cup apparatus and measuring the temperature at regular interval till it reaches the temperature of flash and then the fire point is obtained, and similarly by cooling, the flash and fire points are observed.

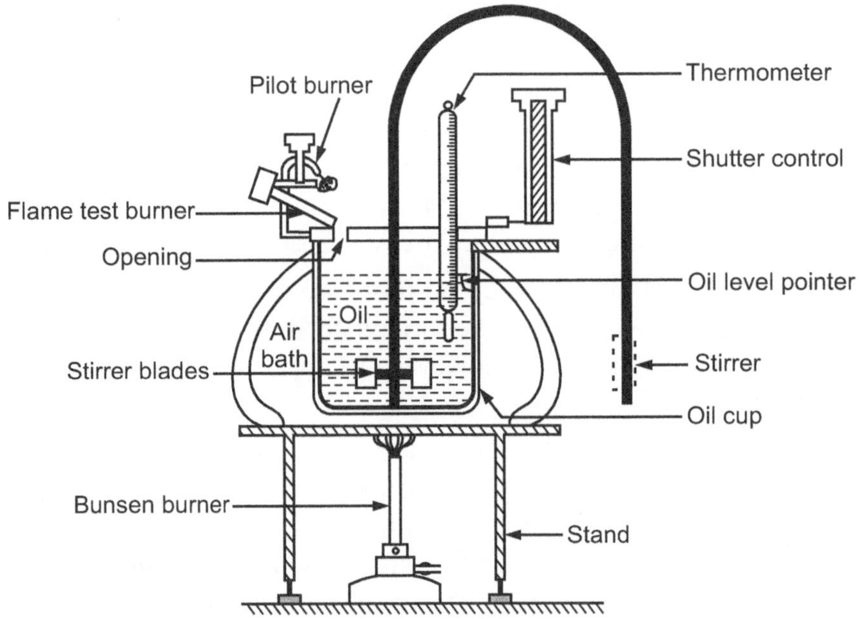

Fig. 2.7: Pensky Martens apparatus (Closed cup)

Clean the cup and fill it with given sample of oil up to the filling mark. Cover the cup with lid. Thermometer is now inserted and should not touch the metallic cup. Heat the oil by means of electric heater or burner so that the sample of oil gives out vapour at the rate of 5°C- 6°C per minute. Stir the sample at one or two revolutions per seconds. When the oil gives out vapour, introduce the test flame above the lid, wait for flash with flickering sound. Introducing test flame should be continued at regular intervals until the first flash is observed with peak flickering sound. The temperature corresponding to this flickering sound is noticed and it is flash point temperature of the given sample of oil. Do not stir the sample while applying the test flame.

Continue the process of heating and introducing the test flame until the oil will begin to burn continuously (for 5 seconds at least) and observe the temperature. This is the fire point temperature of the given sample of oil. Repeat the test twice or thrice with fresh same sample of oil and observe the results.

It is useful for determination of close cup flash points above 49°C. of fuels, oils, other viscous materials and suspension of solid.

2.7.2 Abel's Apparatus

Abel's flash-point apparatus is a petroleum-testing apparatus for determining the flash-point. It is used for the purpose of determining the flammable point of a number of petroleum products. Abel's flash point apparatus with oil test jet is heated by hot plate with digital temperature indicator.

This method of test is designed for the determination of the closed flash points of cutback bitumen's of which the flash points do not exceed 230° F (110° C). Use is made of a modified form of the Abel apparatus. The results obtained by this method do not differ appreciably from the minimum flash point of the material under equilibrium conditions of temperature and vapour pressure.

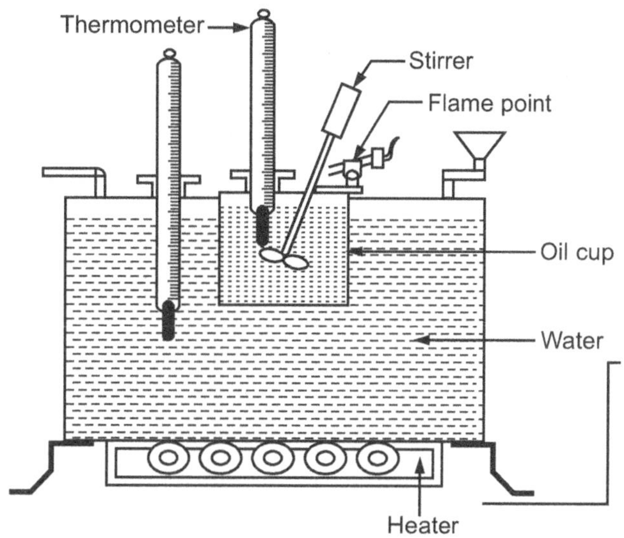

Fig. 2.8: Abel's flash point apparatus set up

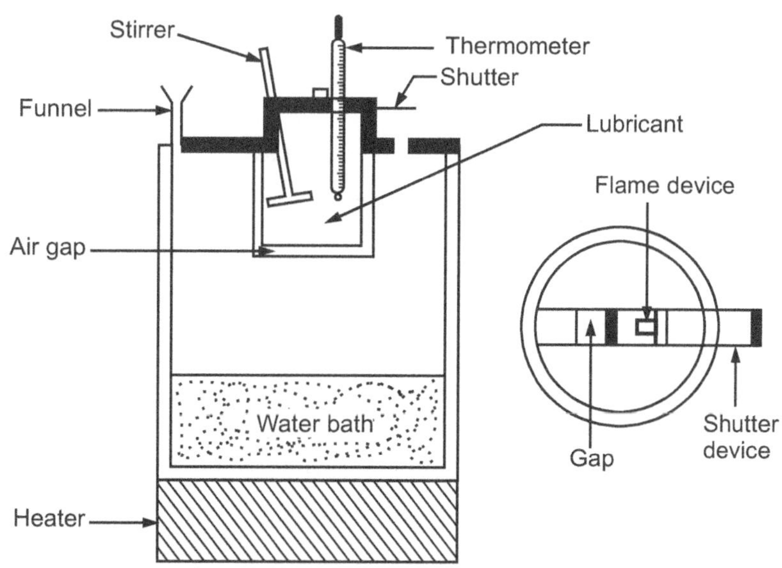

Fig. 2.9: Internal arrangement and top view of Abel's Flash point apparatus

Clean the cup and fill it with given sample of oil up to the filling mark. Insert the thermometer in the holder. Make sure that the thermometer should not touch the metallic cup. Heat the oil by means of electric heater so that the sample of oil gives out vapour at the rate of 10°C per minute. When the oil gives out vapour, introduce the test flame above the oil, without touching the surface of the oil and watch for flash with flickering sound. Introducing test flame should be continued at regular intervals until the first flash is observed with peak flickering sound. The temperature corresponding to this flickering sound is noticed and it is flash point temperature of the given sample of oil. Continue the process of heating and introducing the test flame until the oil will begins to burn continuously and observe the temperature. This is the fire point temperature of the given sample of oil. Repeat the test twice or thrice with fresh same sample of oil and observe the results. The observation is tabulated.

It is suitable for determination of close cup flash point determination of Petroleum and petroleum products having flash points below 70°C.

2.7.3 Cloud and Pour Point Apparatus

These methods determine the cloud and pour point of petroleum products. The method for **Cloud point** (section1) is intended for use only on oils which are transparent in layers 40 mm in thickness and have cloud point below 49°C. The method for **Pour point** (section II) is intended for use on any petroleum oil.

(I) Cloud Point :

The temperature, expressed to the nearest degree Centigrade, at which the oil is cooled under prescribed conditions.

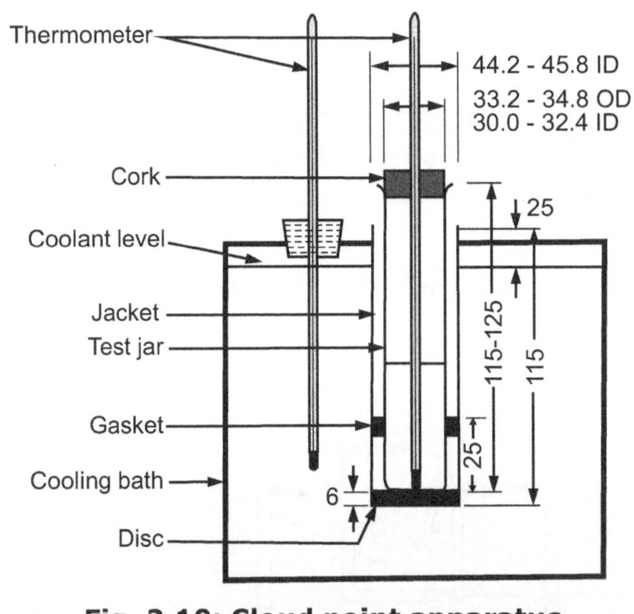

Fig. 2.10: Cloud point apparatus

Bring the sample to a temperature of at least 15°C above the approximate cloud point and pour it into the jar to a height of 51 to 57 mm. Close the jar with the cork so that the thermometer bulb rest on the center of the bottom of the jar. Fit the gasket on to the jar 25 mm from the bottom and insert the jar into the jacket. The disc, the gasket, and jacket shall be kept clean and dry. Maintain the bath temperature at minus 1°C to 2°C. Support the jacket and jar in a vertical position in the bath so that not more than 25 mm projects from the cooling medium. At each thermometer reading of 1°C, remove the jar from the jacket quickly but without disturbing the oil, inspect the material for cloud, and replace the jar; this complete operation shall not take more than 3 second. If the sample does not show a cloud when it has been cooled 10°C, place the jar and jacket in another bath maintained at a temperature of −15°C to −18°C. If the sample does not show a cloud when it has been cooled to −7°C place the jar and jacket in another bath maintained at a temperature of −32°C to −35°C. When an inspection of the sample first reveals a distinct cloudiness or haze at the bottom of the jar, record the reading of the thermometer as the cloud point.

(II) Pour Point:

The lowest temperature, expressed as a multiple of 30°C, at which the oil in observed to flow when cooled and examined under prescribed conditions.

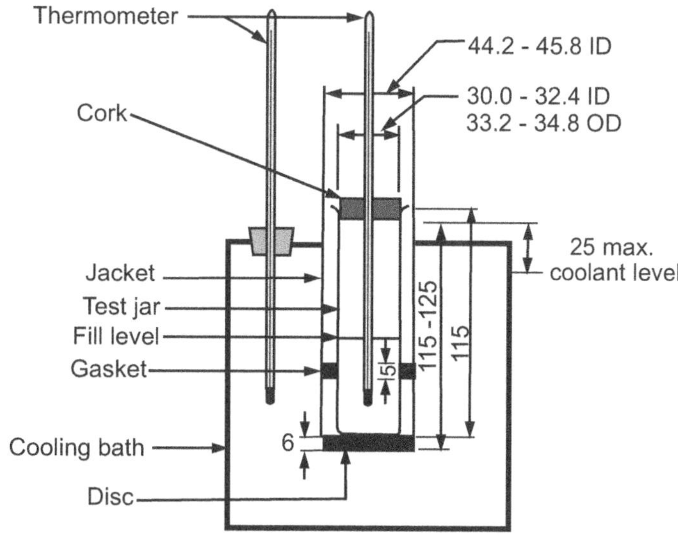

Fig. 2.11: Pour point apparatus

Pour the sample, heated in a water bath if necessary, into the jar to a height of 51 to 57 mm. Close the jar with the cork-carrying thermometer No. 1 so that the thermometer bulb is immersed vertically in the 3 sample with the beginning of the capillary 3 mm below the surface. Heat the sample, without stirring, to a temperature of 46°C in a bath maintained at a temperature not higher than 48°C. Cool the sample to 32°C in air or in a water bath at approximately 25°C. If a pour point below −35°C is expected, cool the sample in air or in water bath to 16°C and replace the thermometer No.1 by thermometer No. 2. Fit the gasket on to the jar 25 mm from the bottom and insert the jar into the jacket. The disc, the gasket, and the jacket shall be kept clean and dry.

If a pour point above 32°C is expected, heat the sample to 46°C or to a temperature 8°C above the expected pour point, and fit the jar into the jacket immediately. One the sample has cooled enough to allow the formation of wax crystals, take great care not to disturb the mass of sample nor to permit the thermometer to shift in the sample. Any disturbance of the spongy network of crystals will lead to false results. Maintain the bath temperature at minus 1°C to plus 2°C. Support the jacket and jar in a vertical position in the bath so that not more than 25 mm projects form the cooling medium. Beginning at a temperature 12°C above the expected pour point, at each thermometer reading which is a multiple of 3°C, remove the jar from the jacket carefully, and tilt it just enough to see whether the oil will move and the replace it, this complete operation shall not take more than 3 seconds. If the oil has not ceased to flow when it has been cooled to 9°C, place the jar and jacket in another bath maintained at a temperature of minus 32°C to minus 35°C. If the pour point is very low, maintain additional baths with successively lower temperature differentials of about 18°C, transfer in the jar and jacket when the temperature of the sample reaches a point 27°C above the temperature of the new bath. As soon as the sample ceases to flow when the jar is tilted, hold the jar in horizontal position for exactly 5 seconds. If the sample shows any movement replace the jar in the jacket and cool down the sample another 3°C. If the oil shows no movement during the 5 second, record the reading of the thermometer.

Certain lubricating oils tend to move as a whole. The point at which they cease to flow is consequently more difficult to detect. For black oil, cylinder stock and non-distillate fuel the result obtained is the upper (maximum) pour point. Determine the lower (Minimum) Pour Point by heating the sample with stirring to 105°C, pouring it into the jar, cooling it to 32°C as before.

2.7.4 Drop Point Apparatus

Dropping point determinations are used for identification and quality control purposes, and can be an indication of the highest temperature of utility for some applications. The sample is heated at a prescribed rate in a precision machined cup whose sides slope toward an opening at its center. The temperature at which a liquid drop first falls from the cup is the dropping point of the sample.

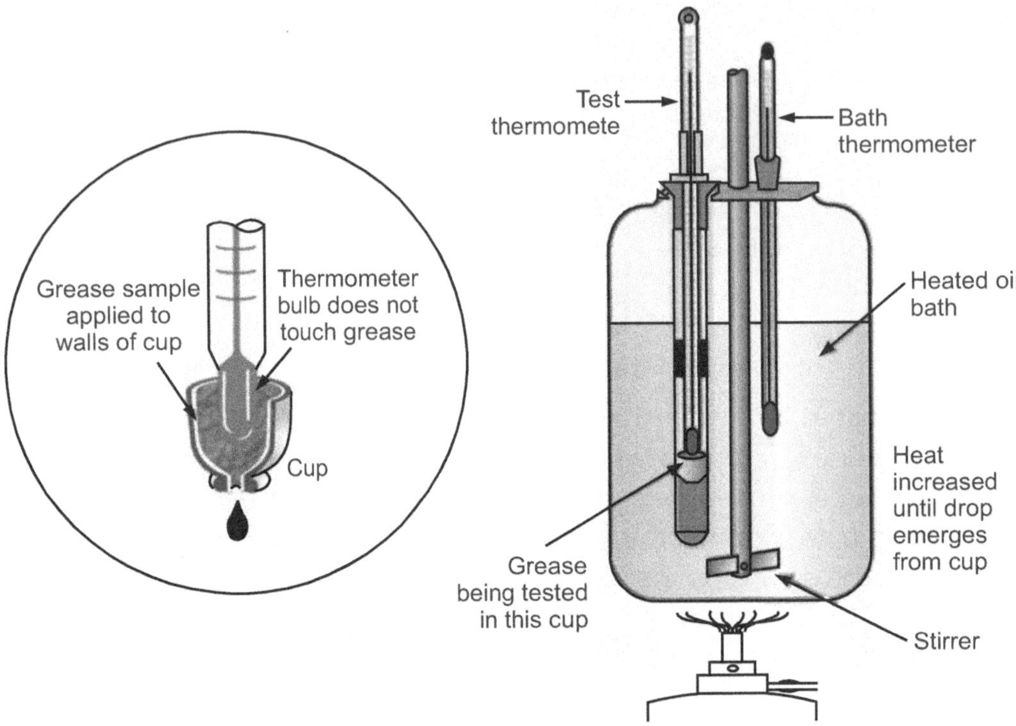

Fig. 2.12: Drop point apparatus

To establish the dropping point, a grease sample is warmed in the testing device until liquid drops fall to the bottom of the test tube through an opening in a nipple. Gel- or powder-based greases, which are measured at temperatures of more than 300°C, are considered as not having a dropping point.

There will not always be a correlation between a grease's dropping point and its maximum operating temperature. Of course, the permissible temperature is always less than the dropping point value. Along with the thickener, the oil type will dictate a grease's maximum operating temperature.

A lower dropping point of a used grease in comparison to a fresh grease may be caused by greases with different thickeners mixing together, the grease containing water or other foreign fluids, or the grease shearing into small particles under extreme loads.

2.7.5 Smoke Point Apparatus

Smoke Point Apparatus determines the smoke point of kerosene and aviation turbine fuel by measuring the maximum flame height prior to smoke formation and comparing the results with those of known standards. Smoke point is an indicator of the combustion qualities of aviation turbine fuels and kerosene. The fuel sample is burned in the Smoke Point Lamp, and the maximum flame height obtainable without smoking is measured.

The smoke point is the maximum flame height in millimeters at which kerosene will burn without smoking, tested under standard conditions, this test method provides an indication of the relative smoke producing properties of Kerosene and aviation turbine fuel in a diffusion flame. The smoke point is related to the hydrocarbon type composition of such fuels.

As the smoke point is high like the paraffinic compounds it will have low smoking tendency (less able to produce smoke). Inversely, the low smoke point like in aromatic compound has high smoking tendency to produce smoke.

Smoking tendency can be calculated as:

Smoking tendency = 320 / Smoking point in mm

2.7.6 Conradson Apparatus

Carbon residue of the fuel is indicative of carbon depositing tendencies, of the fuel at high temperature in an inert atmosphere. Carbon residue for a fuel is measured in weight% or ppm. High carbon residue value is undesirable for a fuel.

When oil is heated to high temperature in absence of sufficient air a portion of the oil will distill, thus leaving a carbon residue. The amount of residue is an index to the extent the oil is likely to decompose in service. Under engine operating conditions, the amount and type of the carbon formed has an important effect upon the performance of engine. While no maximum residue specification for lubricating oils are commonly given, a relatively smaller amount in one oil compared with another should be favourable.

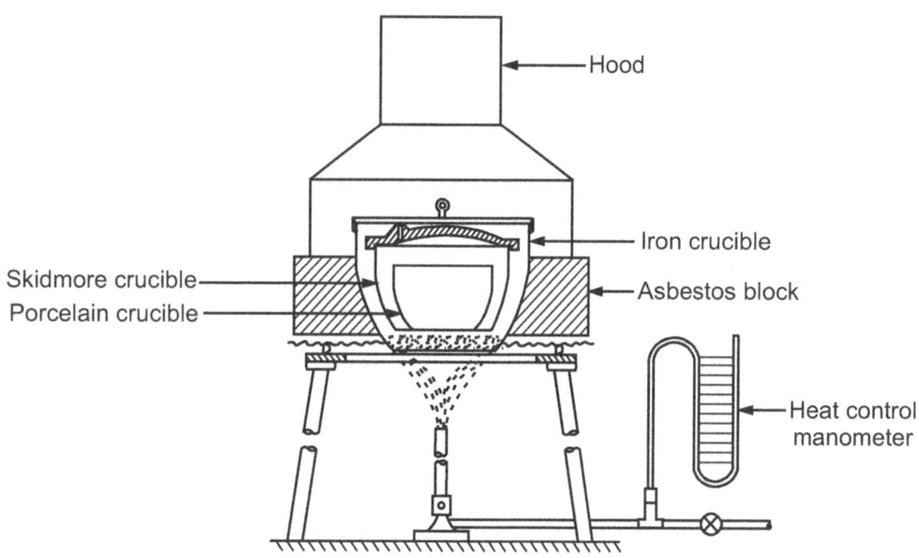

Fig. 2.13: Conradson (Carbon residue) apparatus

It consist of porcelain crucible with a Skidmore iron crucible resting upon a sand in a second larger iron crucible, all of which are supported upon a wire triangle on a metal stand of specified dimensions. The crucibles are surrounded by an asbestos block and covered by a sheet-iron hood and chimney. Heat is supplied by Meker burner. An analytical balance is used to weigh the sample of oil and residue.

2.7.7 Aniline Point Apparatus

The tendency of the lubricating oil to mix with aniline is expressed in terms of Aniline point. Aniline point is the minimum equilibrium solution temperature for equal volumes of aniline and lubricating oil sample.

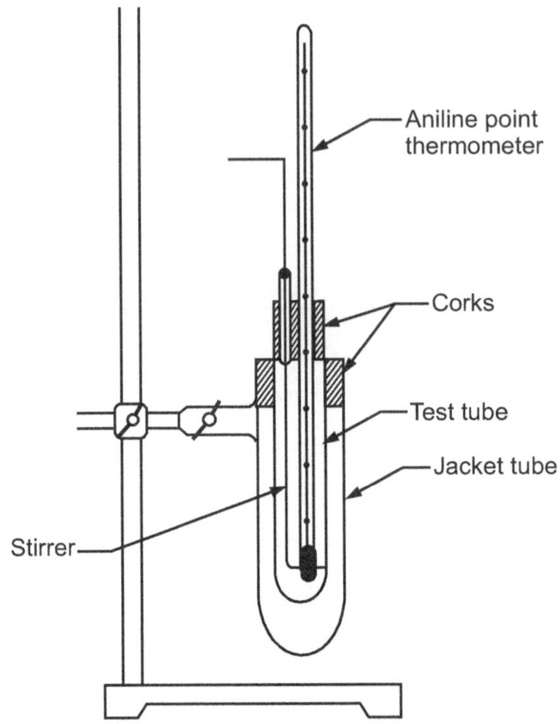

Fig. 2.14: Aniline point apparatus

A lower aniline point of the lubricating oil means a higher percentage of aromatic hydrocarbons in it. Since aromatic hydrocarbons have tendency to dissolve the rubber seals used in the system to prevent leakage. Thus, higher aniline point of lubricating oil is desirable.

$$\text{Aniline point of lubricant} \times \frac{1}{\text{Aromatic contents of lubricant}}$$

The aniline point of the lubricating oil is determined by thoroughly mixing equal volumes of oil sample and aniline in a test tube and heating the mixture on hot bath until a homogenous solution is obtained, then the test tube is allowed to cool at controlled rate. The temperature at which lubricating oil and aniline phases separate out is recorded as the aniline point.

2.8 MEASUREMENT OF PROPERTIES OF PETROLEUM PRODUCTS

Distillation of crude yields number of cuts of varying ranges. These fractions or cuts are processed suitable to utility when they are known as products. In the order of increasing boiling points, the main fraction from distillation of crudes are Gas, Gas oil, Gasoline, Kerosenes, Middle and Heavy distillates, Residuums etc.

2.8.1 Gasoline

Naphtha or Gasoline is the next fraction to follow gas. This is a volatile fraction and is known as motor spirit. The boiling ranges from 37°C to 180°C. Gasoline is a finished product, while raw fraction is termed as naphtha or light boiling fraction. The product is so important that the integration of refineries has become a reality to produce as much as 60-70% of gasoline from crude, especially in countries like U.S.A.

2.8.1.1 ASTM Distillation

It is a important test prescribed for gasoline. In this test, 100 ml of sample is distilled in a standard flask at a uniform rate of 5 cc per minute. The distillate is condensed in a brass tube condenser, surrounded by a water bath kept at 0°C by ice-water mixture. First drop from the condenser must be available in 5 to 10 minutes after heating started, at which the recorded temperature is mentioned as initial boiling point (IBP) of the sample. The vapour temperature is recorded at each successive 10 cc distillate collected in a measuring cylinder. The test continues in the same way till 95% of fraction is condensed. At this juncture, the heat intensity may be increased to obtain the maximum boiling point also known as end point (EP).

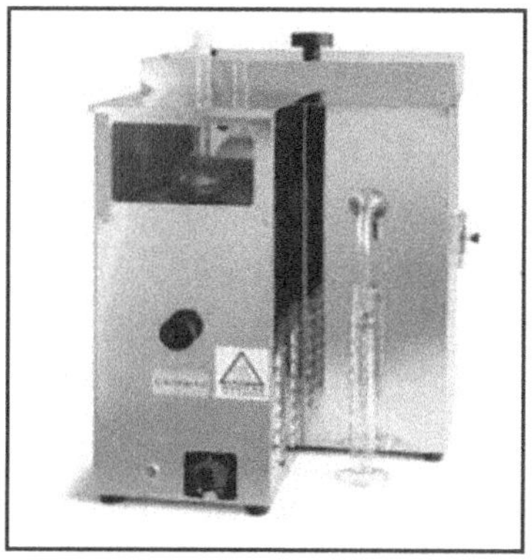

Fig. 2.15: ASTM distillation

Fluctuation in temperature is common when last two or three ccs of sample are distilled. When the bottom of the flask shows dryness, the temperature recorded corresponds to final boiling point. The distillate collected shall not be less than 98% and the difference is accounted as loss; usually ascribed to light ends. This method of distillation is followed for most of the product specifications. When middle distillates like diesel are to be tested, instead of 100 ml sample, a 200 ml sample is taken. Similar adjustment of heating rates are followed to obtain IBP in 10 to 15 minutes time. This test is conducted with standard equipment only. The boiling points in this method need not be adjusted to 760 mm pressure; although it is preferred at times of dispute. ASTM distillation specifies the evaporation characteristics of gasoline. Accordingly gasoline fall mainly into 3 types: Type A, Type B, Type C. In all these types 10%, 50% and 90% boiling point along with IBP and FBP are mentioned.

If the range of boiling point is high, gasoline will not be evaporating normally. If 90% boiling is high, then the crank case dilution also increases as the fuel is not volatile at the working conditions of carburettor.

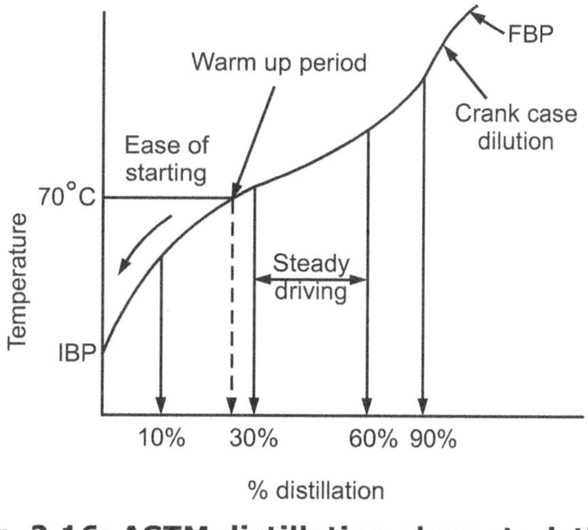

Fig. 2.16: ASTM distillation characteristics

2.8.1.2 Specific Gravity

Specific gravity, also called relative density, is the ratio of the density of a substance to the density of a reference substance; equivalently, it is the ratio of the mass of a substance to the mass of a reference substance for the same given volume. Apparent specific gravity is the ratio of the weight of a volume of the substance to the weight of an equal volume of the reference substance.

Specific gravity varies with temperature and pressure; reference and sample must be compared at the same temperature and pressure or be corrected to a standard reference temperature and pressure. Substances with a specific gravity of 1 are neutrally buoyant in water. Those with Specific gravity greater than 1 are denser than water and will, disregarding surface tension effects, sink in it. Those with a specific gravity less than 1 are less dense than water and will float on it. In scientific work, the relationship of mass to volume is usually expressed directly in terms of the density (mass per unit volume) of the substance under study. It is in industry where specific gravity finds wide application, often for historical reasons.

Specific gravity is commonly used in industry as a simple means of obtaining information about the concentration of solutions of various materials, or of quality control for polymer materials: to evaluate physical changes, or determine the degree of uniformity between samples or lots.

2.8.2 Kerosene

Kerosene is the general name applied to the group of refined petroleum fractions employed as fuel and illuminant. All these fractions have approximate boiling range 150-250°C. These are uniform close cut distillates, low in viscosity, with a good degree of refinement to be fairly stable, light in colour and free from smoky, ill-smelling substances. It is highly desirable that sulphur be kept as low as possible. Kerosene is used as illuminating oil in domestic needs both in wick burner and mantle burners. Such kerosene's were special blends of relatively high boiling fractions with oils of marine animals.

2.8.2.1 Flash Point

This is defined as, "the minimum temperature at which the vapours from oil sample will give a momentary flash on application of a standard flame under specific test conditions". Significance of the flash point is that it can predict the possible fire hazards during transportation, storage and handling. Early tendencies were to incorporate more volatilities in kerosenes, thus inviting unusual fire hazards the flash point of marketable kerosene should be above 42°C.

2.8.2.2 Fire Point

This closely follows the flash point. This test is carried out in open cup rather than a closed one. Fire point temperature is noted when the oil vapours can burn continuously for 5 seconds when tested in flash point apparatus and it occurs after the flash point, by 3-4°C. For domestic needs a very high flash point above 50°C is also not desirable. The easy ignition is credited to volatilities enhances the flash point causing difficulty in ignition.

2.8.2.3 Smoke Point

Smoke point is an indication of clean burning quality of kerosene. Illumination depends upon the flame dimensions although it is not related to flame height. Many paraffins may be gifted with better flame heights but illumination may be poor. Smoke point is defined as, "the maximum height of flame in millimetres at which the given oil will burn without giving smoke". Thus, smoke point test enables the burning quality to be measured by adjusting the wick height to give proper non-smokey flame. Illumination is supposed to be characteristics of the components of fuel usually not measured. Different flame heights are obtained due to presence of different components such as paraffins, naphthenes and aromatics.

Aromatic contribute smoke, hence removal of aromatics increases the smoke point. Naphthenes with side chains are inevitably retained to give good illumination. In india, marketable kerosene should posses a smoke point 18 mm. Smoke point may be predicted approximately by knowing the group composition of oil, given as:

$$\text{Smoke point} = 0.48\ P + 0.32N + 0.20A$$

Where P, N, A are % of paraffins, naphthenes and aromatics in the sample.

High smoke point meant a better quality fuel is not always acceptable as it cannot specify any illumination characteristics.

2.8.2.4 Aniline Point

This test indicates qualitatively the amount of aromatics present in kerosene. This method is fully discussed in diesel fuels.

2.8.3 Diesel

Diesel oils are the fractions in the boiling range of 250-320°C; and fall under gas oil fractions. These are basically divided into two classes as high speed diesels and low speed diesels.

Volatility test for these fuels is of trifle value; however it is advisable to conduct ASTM distillation test to infer the boiling range. High percentage of light ends (upto 10% distillation) are trails of low flash points. Even though flash point has no direct influence upon the performance of fuel, still its minimum value should be maintained to prevent possible fire hazards in handling and storage.

If 10-50% boiling temperature is high, its warm up period will considerably increase. A high boiling point gives much smoke formation and crank case dilution. For all high speed oils 50% point is kept around 300°C; and for low speed 50% point gives a low viscosity and possibly low heating value too.

The general tests recommended for diesel fuels are:

2.8.3.1 Cloud and Pour Point

Cloud point is the temperature at which wax crystals are first observed in the liquid sample upon cooling under prescribed conditions. It is an important physical property for estimating diesel fuel performance under cold temperatures since wax formation can cause flow performance issues such as fuel filter clogging.

The criteria of pour point fixation depends upon two factors namely climatic conditions and storage. Fuel, at minimum ambient temperature must be free flowing. In India pour point is fixed at 5°C; however in the Himalayan belt, where the climate persists at sub zero level this may not be satisfactory; hence low pour point oils are essential. It is also observed that at close approach of pour point (within 2 to 3°C), the viscosity increases very much, the result of which is high pumping cost.

2.8.3.2 Aniline Point

Diesel fuels are mainly composed of paraffins although there is no bar for aromatics. Aromatics of this boiling range present in the fuel cause abnormal ignition delay. For this reason an estimation of aromatics is essential. One such method is Aniline point which can predict the suitability of oil. Aniline point is defined as, "the minimum temperature at which equal volumes of anhydrous aniline and oil mix together". Diesel index a derivative of Aniline point is defined as:

$$[0.018\ \text{A.P}^{\circ}\text{C} + 0.32]\ \text{API}.$$

Aniline being an aromatic compound freely mixes with aromatics; so a low aniline point indicates, low diesel index (because of high percentage of aromatics). Aniline point can also predict the amount of carbon present in the molecule (aromatics), given by the formula

$$\%C_A = 1039.4\, n_D^{20} - 470.4\, d_{20} - 0.567\, AP\,(^\circ C) - 1104.42$$

Where, n_D^{20} is a refractive index at 20°C.

d_{20} is density at 20°C.

Diesel index is a measure of ignition quality of fuel. Diesel engine works on the principle of compression ignition. During compression adiabatically, the air temperature reaches around 600°C, when the fuel in finely atomized form is fed in, it instantaneously explodes. The temperature at which this explosion takes place without the aid of fire is called self ignition temperature. Self ignition temperature is low for paraffins while it is high for aromatics. Thus, a fuel rich in aromatics burns later causing ignition delay and gives rise to what is known as diesel knock. For this reason all diesel fuels are processed to have a diesel index in the range of 45 to 55. A high diesel index is also not desirable, as a fuel rich in aromatics gives rise to better calorific value than paraffin rich fuel for equal weights. When the fuel is injected into the chamber of hot air, first the fuel drops acquire the temperature of surrounding.

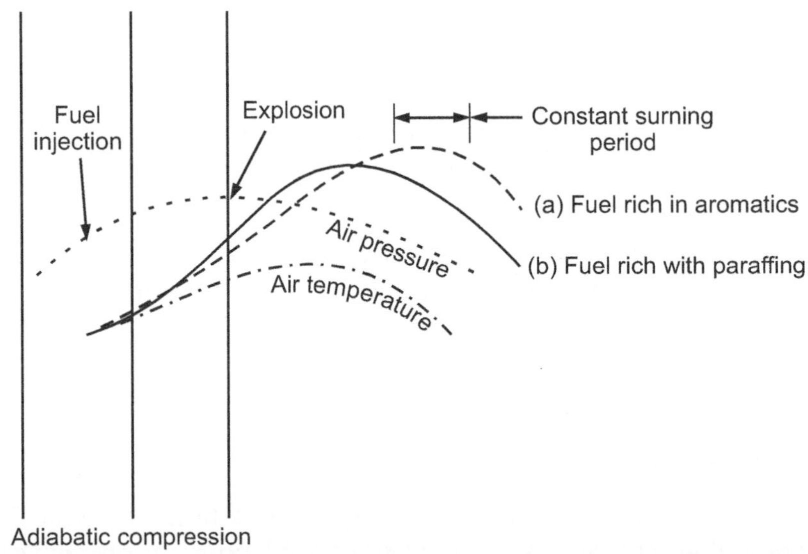

Fig. 2.17: Diesel oil combustion performance in engine

This takes some time. The time delay can be decreased by atomising the fuel into fine drops and also by varying the composition of fuel. Viscosity also has its own role in heat transmission. Once the droplets acquire the temperature of self ignition, droplets explode. A continuous flame front may move if the aromatic rich fuel is fed in-obviously a cause for diesel knocking. Above figure shows the diesel fuel performance in an engine. During adiabatic compression, the temperature rises and at that instant the fuel is injected. The time taken by fuel, from injection to explosion is known as delay period. The fuel burns at a constant rate. The air pressure and temperature increases after explosion. The two curves A & B marked in the figure show the characteristics performance of two fuels - 'A' is for aromatic rich fuel and 'B' is for a fuel rich in paraffins.

Peaks a & b obviously should coincide with crank position, otherwise knocking result. High speed engines require a fuel of viscosity, 1.4 to 2.5 CS (at 30°C) whereas low speed engines may work with even high viscous fuels.

2.8.3.3 Flash Point

It has no real significance on performance of fuel. Knowledge of this is required for safe handling and storing. In India flash point of diesel is kept around 50°C-55°C.

2.8.3.4 Viscosity

Ease of starting depends upon viscosity and ignition quality. At lower starting temperature high cetane fuels are required. Diesels used for different speeds must possess the different properties.

Thermal stability of diesel fuels is important because the remaining fuel from the combustion chamber is returned back to the reservoir. The test is performed by exposing 50 ml sample kept at 149°C for 90 minutes with an additive. When the sample is filtered through a filter paper, the excessive stain on filter paper is an indication of bad oxidation stability of the fuel. Usually additives like 1-n butoxyl-1-ethane, dioctyl-diphenyl amine (upto 20 ppm) are used to improve the stability of fuel.

2.8.4 Wax

Waxes do not go through a sharp solid-liquid phase change when heated and therefore do not have a true melting point. As the temperature rises, waxes gradually soften or become less viscous. For this reason, the determination of the softening point must be made by an arbitrary but closely defined method if test values are to be reproducible. This test is useful in determining the consistency of waxes, and as one element in establishing the uniformity of shipments or source of supply.

2.8.4.1 Drop Point

The drop point is the temperature at which the first drop of molten wax precipitates from a standardized vessel under controlled test conditions. As the temperature rises, waxes gradually soften and melt over a relatively large temperature interval.

This test method has been found suitable for all types of waxes including paraffin, microcrystalline polyethylene, and natural waxes.

2.8.5 Lubricating Oil

The principal source of lubricating oil is the fraction that is left after lighter components, namely gasoline, kerosene, diesel oil during crude distillation. Generally, lubes have a boiling point above 350°C and these are obtained as a main products from vacuum distillation units. Residuum's, after precipitation of asphaltenes are known as bright stocks and form a good source for lube oils.

Important Points

- A refinery is composed of a group of chemical engineering unit processes and unit operations used for refining certain materials or converting raw material into products of value.
- The various types of refineries are :
  - **Oil refinery :** Converts petroleum crude oil into high-octane motor fuel (gasoline/petrol), diesel oil, liquefied petroleum gases (LPG), jet aircraft fuel, kerosene, heating fuel oils, lubricating oils, asphalt and petroleum coke.
  - **Sugar refinery :** Converts sugar cane and sugar beets into crystallized sugar and sugar syrups.

- **Natural gas processing plant :** Purifies and converts raw natural gas into residential, commercial and industrial fuel gas and also recovers natural gas liquids (NGL), such as ethane, propane, butanes and pentanes.
- **Salt refinery :** Cleans salt (NaCl), produced by the solar evaporation of sea water, followed by washing and re-crystallization.

- **Refining Process :** Refineries combine the various fractions (processed, unprocessed) into mixtures to make desired products. Petroleum refining is the physical, thermal and chemical separation of crude oil into its major distillation fractions which are then further processed through a series of separation and conversion steps into finished petroleum products.

  About 90% of the petroleum products used in the U.S. are fuels with motor gasoline accounting for about 43% of the total.

- **Unit Operations in Refining Process :** It is a basic step in a process. It also fall in the following categories :
  - Combination (mixing)
  - Separation (distillation)
  - Reaction (chemical reaction)

- **Major Unit Operations in Refining Process :**

  1. **Desalting :** This treatment is prior to distillation. It removes some of the metals and suspended solids which cause catalyst deactivation. By this process we have to reduce corrosion, plugging and fouling of equipment and to prevent poisoning the catalysts in processing units.
  2. **Crude Oil Distillation (Fractionation) :** Fractional distillation is useful for separating a mixture of substance with narrow difference in boiling points, and is the most important step in the refining process. It is further divided into :

     (a) Atmospheric Distillation Tower : It is a vertical column with perforated trays with bubble casps. At successively higher temperature on the top of tower, the various major products including lubricating oil, heating oil, kerosene, gasoline and uncondensed gases (which condense at lower temperatures) are drawn off.

     (b) Vacuum Distillation Tower : This vacuum distillation process has become an important chain in maximizing the upgrading of crude oil. It is simply the distillation of petroleum fractions at a very low pressure (0.2 to 0.7 psi) to increase volatilization and separation.
  3. **Solvent Extraction :** This process separates aromatics, naphthenes, and impurities from the product stream by dissolving or precipitation. It separates aromatics, naphthenes and impurities from the product stream by dissolving or precipitation.
  4. **Absorption Process :** This process selectively removes a certain gas from a gas mixture using a liquid absorbent. In the refining industry, this process is used extensively to free the product gas streams from acid gases (mainly H_2S) either by using a physical or a chemical absorbent.
  5. **Adsorption Process :** Adsorption processes use a solid material (adsorbent) possessing a large surface area and the ability to selectively adsorb a gas or a liquid on its surface. The operation conditions are approximately 370°C and 100 psi.

     Example : Silica (SiO_2), Anhydrous alumina (Al_2O_3) and Molecular sieves (crystalline silica/alumina).

- **Petroleum Hyrocarbon Structures :** It consists of thee main hydrocarbon groups.
  1. **Paraffins :** It consists of straight or branched carbon rings saturated with hydrogen atoms, the simplest of which is methane (CH_4) the main ingredient of natural gas. Others in this group include ethane (C_2H_6) and propane (C_3H_8).
  2. **Naphthenes :** It consist of carbon rings, sometimes with side chains, saturated with hydrogen atoms. Naphthenes are chemically stable, they occur naturally in crude oil and have properties similar to paraffins.
  3. **Aromatics :** Aromatic hydrocarbons are compounds that contain a ring of six carbon atoms with alternating double and single bonds and six attached hydrogen atoms. This type of structure is known as a *Benzene Ring*.
- **Petroleum Products :** Petroleum products are any oil-based products which can be obtained by distillation and are normally used outside the reining industry.

  The various fractions obtained are as follows :
  1. **Aviation Gasoline :** It have an octane number suited to the engine, a freezing point of −60°C, and a distillation range usually within the limits of 30°C and 180°C.
  2. **Gas Diesel Oil/(Distillate Fuel Oil) :** It distills between 180°C and 380°C. Several grades are available depending on uses.
  3. **Heavy Fuel Oil Residual :** Its kinematic viscosity is above 10 CST at 80°C. The flash point is always above 50°C and the density is always more than 900 kg/lit.
  4. **Kerosene :** It is a medium oil, distilling between 150°C and 300°C. A special cut of kerosene can be processed for the manufacture of Linear Alkyl Benzene used as a detergent.
  5. **Jet Fuel :** This category comprises both gasoline and kerosene type jet fuels meeting specifications for use in aviation turbine power units.
     - **(a) Gasoline type jet fuel :** They distill between 100°C and 250°C. It is obtained by blending kerosenes and gasoline or naphthas.
     - **(b) Kerosene type jet fuel :** This is a medium distillate used for aviation turbine power units. It has the same distillation characteristics and flash point as kerosene.
  6. **LPG (Liquefied Petroleum Gas) :** It is a colourless paraffinic gas which is extracted from natural gas and refinery gas streams. It is generally a mixture of propane and butane. At atmospheric temperature and pressure it is gas. It has therefore to be stored under pressure to keep it liquefied, hence the name LPG.
  7. **Motor Gasoline :** It is distilled between 35°C and 215°C and is used as a fuel for land based spark ignition engines. This is used in internal combustion engines such as motor vehicles, excluding aircraft.
  8. **Naphtha :** Naphtha comprises material in the 30°C and 210°C distillation range or part of this range. The most important usage of naphtha is in the production of Motor Spirit. Naphtha is also an important feedstock for fertilizer manufacture.
  9. **Other Petroleum Products :**

     White Spirit and SBP : White spirit and SBP are refined distillate intermediates with a distillation in the naphtha/kerosene range. They are sub-divided as :
     - **(a) Industrial spirit (SBP) :** SBP is a light oil of narrower cut than motor spirit. There are 7 or 8 grades of industrial spirit.
     - **(b) White spirit :** Industrial spirit with a flash point above 30°C. The distillation range of white spirit is 135°C to 200°C.
     - **(c) Lubricants :** They are used to reduce friction between bearing surfaces. This category includes all finished grades of lubricating oil, from spindle oil to cylinder oil.

(d) **Bitumen :** They are black in colour, obtained as a residue in the distillation of crude oil. It is used for surfacing of roads and for roofing material.

(e) **Paraffin waxes :** They are colourless, odourless and translucent, with a melting point above 45°C.

(f) **Others :** Includes the petroleum products not classified above, for example : tar, sulphur and grease.

10. **Petroleum Coke :** It is used as a feedstock in coke ovens for the steel industry, for heating purposes, for electrode manufacture and for production of chemicals.
11. **Refinery Gas :** It consist mainly of hydrogen, methane, ethane and olefins. It also includes gases which are returned from the petrochemical industry.
12. **A.T.F. (Aviation Turbine Fuel) :** It is also called the jet fuel which having a specifications such as the freezing point of ATF should be below 47°C.
13. **HSD (High Speed Diesel) :** It is an important fuel used for road, rail and marine transport systems.
14. **LDO (Light Diesel Oil) :** It is used as fuel in low speed engines.
15. **FO (Fuel Oil) :** This is used as a fuel in furnaces.
16. **Bitumen :** They are used for road making.

Practice Questions

1. Explain Refining process in detail.
2. Explain Crude oil pre-treatment process.
3. Explain Petroleum hydrocarbon structure.
4. State and explain the different products of crude oil with their temperature range and carbon atoms content.

MSBTE Questions and Answers (As Per G-Scheme)

Summer 2016

1. Explain desalting of crude. **(4 M)**

Ans. Refer Section 2.3.1.

2. Name various fractions obtained from petroleum distillation. Write their approximate boiling range. **(4 M)**

Ans. Refer Section 2.5.

3. What do you mean by LEL and HEL? What is ignition temperature? **(4 M)**

Ans. **LEL:** Lower Explosive Limit (LEL) is the minimum concentration of a particular combustible gas or vapour necessary to support its combustion in air. Below this level, the mixture does not burn.

HEL: The maximum concentration of a gas or vapour that will burn in air is defined as Higher Explosive Limit (HEL). Above this, the mixture is too rich to burn.

Ignition temperature: Refer Section 2.6 (13).

4. Define: **(4 M)**

(i) Flash point.

Ans. Refer Section 2.6 (1).

(ii) Drop point.

Ans. Refer Section 2.6 (9).

(iii) Pour point.

Ans. Refer Section 2.6 (8).

(iv) Cloud point.

Ans. Refer Section 2.6 (7).

5. Why distillation operation is considered to be major unit operation in oil refining process? **(4 M)**

Ans. Refer Section 2.4.1 (I).

6. How crude oil is separated by fractional distillation? **(4 M)**

Ans. Refer Section 2.4.

7. What is the importance of vacuum distillation in refinery operations. **(4 M)**

Ans. Refer Section 2.4 (II).

8. List various types of refineries. **(4 M)**

Ans. Refer Section 2.1.

9. Define: **(4 M)**

(i) Octane number.

(ii) Cetane number.

Ans. Refer Section 2.6 (12) and (11) respectively.

10. Describe the working of vacuum distillation unit. **(4 M)**

Ans. Refer Section 2.4 (II).

Summer 2017

1. Name any four test properties of each gasoline and diesel. **(4 M)**

Ans. Refer Section 2.6.

2. Write the boiling range of any four fractions obtained from fractionation of crude oil. **(4 M)**

Ans. Refer Table 2.5.

3. Write the carbon range of any four fractions obtained from fractionation of crude oil. **(4 M)**

Ans. Refer Section 2.5.

4. Describe refining process with well labelled flow diagram. **(4 M)**

Ans. Refer Section 2.1.1.

5. Define following: **(4 M)**

(i) Octane number

(ii) Cetane number

Ans. Refer Section 2.6 (12) and (11) respectively.

6. Name any two test properties of the following:

(i) Gasoline, (ii) Kerosene, (iii) Diesel, (iv) Lubricating oil. **(4 M)**

Ans. Refer Section 2.6.

7. Explain with well labelled flow diagram, atmospheric distillation unit. **(4 M)**

Ans. Refer Section 2.4.

8. Name four methods for dehydration and desaltification of crude oil. Explain any one in brief. **(4 M)**

Ans. Refer Section 2.3.

9. Write any one use of the following : **(4 M)**

(i) Kerosene

Ans. Refer Section 2.5 (4).

(ii) Diesel

Ans. Refer Section 2.5 (2)

(iii) Lubricating oil

Ans. Refer Section 2.5.

(iv) Gasoline

Ans. Refer Section 2.5 (1).

10 Draw well labelled diagram of vacuum distillation unit. **(4 M)**

Ans. Refer 2.4 (II).

11. Name the fraction which belongs to following list of test properties.
(i) Octane number, (ii) Aniline point, (iii) Smoke point, (iv) Cetane number. **(4 M)**

Ans. Refer Section 2.6.

Winter 2017

1. Define octane number and calorific value. **(4 M)**

Ans. Refer Section 2.6 (12 and 14).

2. What are the various properties of crude oil ? (Any four) **(4 M)**

Ans. Refer Section 2.6.

3. List various fractions obtained in distillation of crude oil with their boiling ranges. **(4 M)**

Ans. Refer Section 2.5.

4. What is desalting of crude oil ? Explain with diagram. **(4 M)**

Ans. Refer Section 2.3.

5. State the uses of hydrocarbons/fractions obtained from petroleum distillation. **(4 M)**

Ans. Refer Section 2.5.

6. Name and explain unit operations in separation process of refining of petroleum. (Any four) **(4 M)**

Ans. Refer Section 2.3.

7. Why separation of crude oil is carried in vacuum distillation? Why low sulphur crude is more preferred? **(4 M)**

Ans. Refer Section 2.4 (II).

8. Explain atmospheric distillation process with neat diagram. **(4 M)**

Ans. Refer Section 2.4 (I).

Summer 2018

1. Explain desalting of crude oil. **(4 M)**

Ans. Refer Section 2.3.

2. Explain the crude oil refining by atmospheric distillation. **(4 M)**

Ans. Refer Section 2.4.1. (I).

3. Define the following :
(i) Octane number. (ii) Cetane number. **(4 M)**

Ans. Refer Section 2.6 (12) and (11) respectively.

4. Define the following term :
(i) Aniline point. (ii) Power point. **(4 M)**
(iii)Cloud point. (iv) Fire point.

Ans. Refer Section 2.6.

5. What is the importance of vacuum distillation in petroleum refining? **(4 M)**

Ans. Refer Section 2.3.2 (II).

6. List various fractions obtained in distillation of crude oil with their uses and boiling ranges.

Ans. Refer Section 2.5. **(8 M)**

✍✍✍

Chapter **3...**

PROCESSES IN REFINERY

Weightage of Marks = 18, Teaching Hours = 12

Syllabus

3.1 Cracking: Definition, Types of cracking (thermal and catalytic).

3.2 Thermal cracking: Visbreaking and delayed coking. (Flow sheet and reaction of following).

3.3 Catalytic cracking: Fixed and fluidized bed catalytic.

3.4 Reforming: Catalytic reforming (platforming).

3.5 Hydrogenation, Hydrocracking and Isomerization.

3.6 Alkylation: Sulfuric acid and (Hydrofloric acid) HF alkylation process.

3.7 Esterification and Hydration.

3.8 Waste treatment: Oil removal, Organic content removal, Solid and hazardous waste treatment.

About this Chapter

At the end of this chapter, students will be able to:

- Explain with sketches the specified type of cracking process.
- Compare the specified refinery processes based on the given criteria.
- Suggest upgradation process for the petroleum product with justification.
- Describe with sketches the specified treatment of the waste water from oil refinery.

3.1 INTRODUCTION

An oil refinery or petroleum refinery is an industrial process plant where crude oil is processed and refined into more useful petroleum products, such as gasoline, diesel fuel, asphalt base, heating oil, kerosene, and liquefied petroleum gas. Oil refineries are typically large sprawling industrial complexes with extensive piping running throughout, carrying streams of fluids between large chemical processing units. In many ways, oil refineries use much of the technology of, and can be thought of as types of chemical plants. The crude oil feedstock has typically been processed by an oil production plant. There is usually an oil depot (tank farm) at or near an oil refinery for storage of bulk liquid products. An oil refinery is considered as an essential part of the downstream side of the petroleum industry.

Petroleum refining processes are those chemical engineering processes and other facilities used in petroleum refineries to transform crude oil into useful products such as liquefied petroleum gas (LPG), gasoline or petrol, kerosene, jet fuel, diesel oil and fuel oils.

Petroleum refineries are very large industrial complexes that involve many different processing units and auxiliary facilities such as utility units and storage tanks. Each refinery has its own unique arrangement and combination of refining processes largely determined by the refinery location, desired products and economic considerations. There are most probably no two refineries that are identical in every respect. Some modern petroleum refineries process as much as 800,000 to 900,000 barrels (127,000 to 143,000 cubic metres) per day of crude oil.

3.2 UNIT PROCESSES IN REFINERIES

3.2.1 Hydrogenation [S-17, 18]

Hydrogenation is a class of chemical reactions which result in an addition of hydrogen (H_2) usually to unsaturated organic compounds. Typical substrates include alkenes, alkynes, ketones, nitriles and imines. Most hydrogenations involve the direct addition of diatomic hydrogen (H_2) but some involve the alternative sources of hydrogen. The reverse reaction, removal of hydrogen, is called **dehydrogenation**.

The classical example of hydrogenation is the addition of hydrogen on unsaturated bonds between carbon atoms, converting alkenes to alkanes. Numerous important applications are found in the petrochemical, pharmaceutical and food industries. Health concerns associated with the hydrogenation of unsaturated fats to produce saturated fats and trans fats is an important aspect of current consumer awareness.

Process :

Hydrogenation has three components: the unsaturated substrate, the hydrogen (or hydrogen source) and, invariably, a catalyst. The largest scale technological uses of H_2 are the hydrogenation and hydrogenolysis reactions associated with both heavy and fine chemical industries. *Hydrogenation* is the addition of H_2 to unsaturated organic compounds such as alkenes to give alkanes and aldehydes to give alcohols. *Hydrogenolysis* is the cleavage of C-X (X = O, S, N) bonds by H_2 to give C-H and H-X bonds. Large-scale applications of hydrogenolysis reactions are associated with the upgrading of fossil fuels. Hydrogenation and hydrogenolysis reactions require metal catalysts, often those composed of platinum or similar precious metals. It is a curious fact that under mild conditions, H_2 reacts directly with organic compound in the absence of such catalysts.

The addition of H_2 to an alkene affords an alkane in the protypical reaction:

$$RCH = CH_2 + H_2 \rightarrow RCH_2CH_3 \text{ (R = alkyl, aryl)}$$

Catalysts :

With rare exception, no reaction below 480°C occurs between H_2 and organic compounds in the absence of metal catalysts. The catalyst simultaneously binds both the H_2 and the unsaturated substrate and facilitates their union. Platinum group metals, particularly platinum, palladium, rhodium and ruthenium, are highly active catalysts. Highly active catalysts operate at lower temperatures and lower pressures of H_2. Non-precious metal catalysts, especially those based on nickel have also been developed as economical alternatives but they are often slower or require higher temperatures. The trade-off is activity (speed of reaction) vs. cost of the catalyst and cost of the apparatus required for use of high pressures.

Two broad families of catalysts are known - homogeneous and heterogeneous. *Homogeneous catalysts* dissolve in the solvent that contains the unsaturated substrate. *Heterogeneous catalysts* are solids that are suspended in the same solvent with the substrate or are treated with gaseous substrate. In the pharmaceutical industry and for special chemical applications, soluble "homogeneous" catalyst are sometimes employed, such as the rhodium-based compound known as Wilkinson's catalyst, or the iridium-based Crabtree's catalyst.

The activity and selectivity of catalysts can be adjusted by changing the environment around the metal, i.e. the co-ordination sphere. Different faces of a crystalline heterogeneous catalyst display distinct activities. Similarly, heterogeneous catalysts are affected by their supports, i.e. the material with the heterogeneous catalyst is bound. Homogeneous catalysts are affected by their ligands. In many cases, highly empirical modifications involve selective "poisons". Thus, a carefully chosen catalyst can be used to hydrogenate some functional groups without affecting others, such as the hydrogenation of alkenes without touching aromatic rings, or the selective hydrogenation of alkynes to alkenes using Lindlar's catalyst. For prochiral substrates, the selectivity of the catalyst can be adjusted such that one enantiomeric product is produced.

Hydrogenolysis :

The catalytic hydrogenation of organic sulfur compounds to form gaseous hydrogen sulfide (H_2S) is very widely used in petroleum refineries, petrochemical plants and other industries to desulfurize various final products, intermediate products and process feedstocks by converting sulfur compounds to gaseous hydrogen sulfide which is then easily removed by distillation. The gaseous hydrogen sulfide is subsequently recovered in an amine treater and finally converted to elemental sulfur in a Claus process unit. In these industries, desulfurization process units are often referred to as hydrodesulfurizers (HDS) or hydrotreaters (HDT). In the petroleum refining and petrochemical industries, cobalt-molybdenum or nickel-molybdenum catalysts are commonly used for hydrogenation and hydrogenolysis catalysts.

3.2.2 Catalytic Cracking [S-14, 16, 17]

Introduction :

- It's principal aim is to crack **lower-value stocks** and produce **higher-value light and middle distillates.** The process also produces light hydrocarbon gases, which are important feed stocks for petrochemicals.
- Catalytic cracking produces more gasoline of higher octane than thermal cracking.
- Products from catalytic cracking units are also more stable due to a lower olefin content in the liquid products. This reflects a higher hydrogen transfer activity, which leads to more saturated hydrocarbons than in thermally cracked products from delayed coking units.
- A major difference between thermal and catalytic cracking is that, reactions through catalytic cracking occur via **carbocation intermediate**, compared to the free radical intermediate in thermal cracking.
- Carbocations are longer lived and accordingly more selective than free radicals. Acid catalysts such as amorphous silica-alumina and crystalline zeolites promote the formation of carbocations.

Description :

(a) Catalytic cracking breaks complex hydrocarbons into simpler molecules in order to increase the quality and quantity of lighter, more desirable products and decrease the amount of residuals. This process rearranges the molecular structure of hydrocarbon compounds to convert heavy hydrocarbon feedstock into lighter fractions such as kerosene, gasoline, LPG, heating oil and petrochemical feedstock.

(b) Catalytic cracking is similar to thermal cracking except that catalysts facilitate the conversion of the heavier molecules into lighter products. Use of a catalyst (a material that assists a chemical reaction but does not take part in it) in the cracking reaction increases the yield of improved-quality products under much less severe operating conditions than in thermal cracking. Typical temperatures are from 450°-510°C at much lower pressures of 10-20 psi. The catalysts used in refinery cracking units are typically solid materials (zeolite, aluminum hydrosilicate, treated bentonite clay, fuller's earth, bauxite and silica-alumina) that come in the form of powders, beads, pellets or shaped materials called *extrudites*.

(c) There are three basic functions in the catalytic cracking process :

- **Reaction :** Feedstock reacts with catalyst and cracks into different hydrocarbons;
- **Regeneration :** Catalyst is reactivated by burning of coke; and
- **Fractionation :** Cracked hydrocarbon stream is separated into various products.

(d) Catalytic cracking uses heat, pressure and a catalyst to break larger hydrocarbon molecules into smaller, lighter molecules. Feed stocks are light and heavy oils from the crude oil distillation unit which are processed primarily into gasoline as well as some fuel oil and light gases. Most catalysts used in catalytic cracking consist of mixtures of crystalline synthetic **silica-alumina** and **amorphous synthetic silica-alumina**. The catalytic cracking processes, as well as most other refinery catalytic processes, produce coke which collects on the catalyst surface and diminishes its catalytic properties. The catalyst, therefore, needs to be regenerated continuously or periodically essentially by burning the coke off the catalyst at high temperatures. The method and frequency in which catalysts are regenerated are major factors in the design of catalytic cracking units.

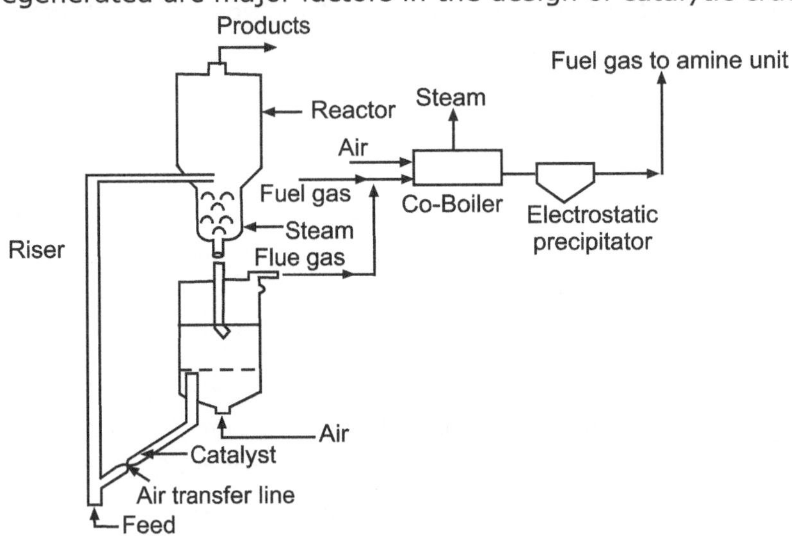

Fig. 3.1: Catalytic cracking flow diagram

Table: 3.1: Catalytic cracking process

Feedstock	From	Process	Typical products	To
Gas oils	Towers, coker	Decomposition, alteration	Gasoline	Treater or blend
	Visbreaker		Gases	Gas plant
			Middle distillates	Hydrotreat, blend, or recycle
Deasphalted oils	Deasphalter		Petrochem feedstock	Petrochem or other
			Residue	Residual fuel blend

(e) The three types of catalytic cracking processes are fluid catalytic cracking (FCC), moving-bed catalytic cracking, and Thermofor catalytic cracking (TCC). The catalytic cracking process is very flexible, and operating parameters can be adjusted to meet changing product demand. In addition to cracking, catalytic activities include dehydrogenation, hydrogenation and isomerization.

3.2.3 Fluid Catalytic Cracking

Description :

(a) The most common process is FCC, in which the oil is cracked in the presence of a finely divided catalyst which is maintained in an aerated or fluidized state by the oil vapours. The fluid cracker consists of a catalyst section and a fractionating section that operate together as an integrated processing unit. The catalyst section contains the reactor and regenerator, which, with the standpipe and riser, forms the catalyst circulation unit. The fluid catalyst is continuously circulated between the reactor and the regenerator using air, oil vapours, and steam as the conveying media.

(b) A typical FCC process involves mixing of a preheated hydrocarbon charge with hot, regenerated catalyst as it enters the riser leading to the reactor. The charge is combined with a recycle stream within the riser, vapourized and raised to reactor temperature (480°-540°C) by the hot catalyst. As the mixture travels up the riser, the charge is cracked at 10-30 psi. In the more modern FCC units, all cracking takes place in the riser. The "reactor" no longer functions as a reactor; it merely serves as a holding vessel for the cyclones. This cracking continues until the oil vapours are separated from the catalyst in the reactor cyclones. The resultant product stream (cracked product) is then charged to a fractionating column where it is separated into fractions, and some of the heavy oil is recycled to the riser.

(c) Spent catalyst is regenerated to get rid of coke that collects on the catalyst during the process. Spent catalyst flows through the catalyst stripper to the regenerator, where most of the coke deposits burn-off at the bottom where preheated air and spent catalyst are mixed. Fresh catalyst is added and worn-out catalyst is removed to optimize the cracking process.

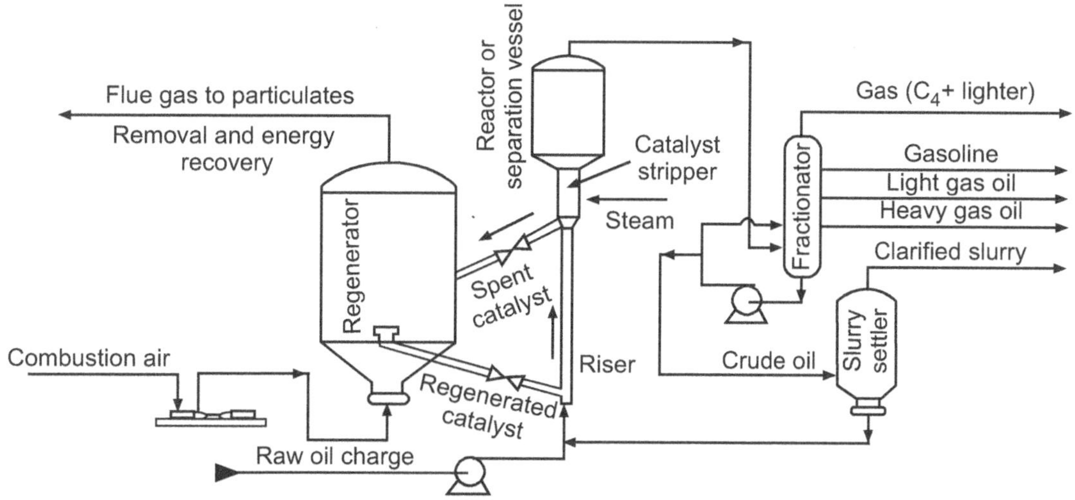

Fig. 3.2 : Fluid catalytic cracking

3.2.3.1 Moving Bed Catalytic Cracking **[S-14; W-17]**

The moving bed catalytic cracking process is similar to the FCC process. The catalyst is in the form of pellets that are moved continuously to the top of the unit by conveyor or pneumatic lift tubes to a storage hopper, then flow downward by gravity through the reactor, and finally to a regenerator. The regenerator and hopper are isolated from the reactor by steam seals. The cracked product is separated into recycle gas, oil, clarified oil, distillate, naphtha and wet gas.

3.2.3.2 Thermofor Catalytic Cracking

In a typical thermofor catalytic cracking unit, the preheated feedstock flows by gravity through the catalytic reactor bed. The vapours are separated from the catalyst and sent to a fractionating tower. The spent catalyst is regenerated, cooled and recycled. The flue gas from regeneration is sent to a carbon-monoxide boiler for heat recovery.

3.2.4 Thermal Cracking

[W-13, 17; S-14, 17, 18]

Description :

(a) Because the simple distillation of crude oil produces amounts and types of products that are not consistent with those required by the market place, subsequent refinery processes change the product mix by altering the molecular structure of the hydrocarbons. One of the ways of accomplishing this change is through "cracking", a process that breaks or cracks the heavier, higher boiling-point petroleum fractions into more valuable products such as gasoline, fuel oil and gas oils. The two basic types of cracking are thermal cracking, using heat and pressure, and catalytic cracking.

(b) The first thermal cracking process was developed around 1913. In this, distillate fuels and heavy oils were heated under pressure in large drums until they are cracked into smaller molecules with better antiknock characteristics. However, this method produce large amounts of solid, unwanted coke. This early process has evolved into the following applications of thermal cracking: visbreaking, steam cracking and coking.

Visbreaking Process :

Visbreaking, a mild form of thermal cracking, significantly lowers the viscosity of heavy crude-oil residue without affecting the boiling point range. Residual from the atmospheric distillation tower is heated (425°-510°C) at atmospheric pressure and mildly cracked in a heater. It is then quenched with cool gas oil to control over-cracking, and flashed in a distillation tower. Visbreaking is used to reduce the pour point of waxy residues and reduce the viscosity of residues used for blending with lighter fuel oils. Middle distillates may also be produced, depending on product demand. The thermally cracked residue tar, which accumulates in the bottom of the fractionation tower, is vacuum flashed in a stripper and the distillate recycled.

Table 3.2 : Visbreaking process

Feedstock	From	Process	Typical products	To
Residual	Atmospheric tower and Vacuum water	Decompose	Gasoline or distillate	Hydrotreating
			Vapour	Hydrotreater
			Residue	Stripper or recycle
			Gases	Gas plant

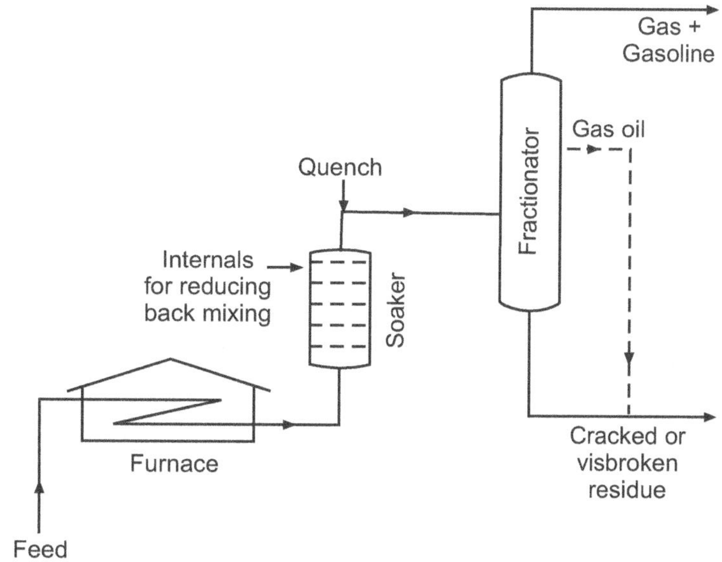

Fig. 3.3 : Visbreaking

Steam Cracking Process :

Steam cracking is a petrochemical process sometimes used in refineries to produce olefinic raw materials (e.g., ethylene) from various feedstock for petrochemicals manufacture. The feedstock range from ethane to vacuum gas oil, with heavier feeds giving higher yields of by-products such as naphtha. The most common feeds are ethane, butane and naphtha. Steam cracking is carried out at temperatures of 800°-870°C, and at pressures slightly above atmospheric. Naphtha produced from steam cracking contains benzene, which is extracted prior to hydrotreating. Residual from steam cracking is sometimes blended into heavy fuels.

Coking Processes : [S-14, 16; W-13]

Coking is a severe method of thermal cracking used to upgrade heavy residuals into lighter products or distillates. Coking produces straight-run gasoline (coker naphtha) and various middle-distillate fractions are used as catalytic cracking feedstock. The process so completely reduces hydrogen so that the residue is a form of carbon called "coke". The two most common processes are delayed coking and continuous (contact or fluid) coking. Three typical types of coke are obtained (sponge coke, honeycomb coke and needle coke) depending upon the reaction mechanism, time, temperature and the crude feedstock.

(a) Delayed Coking : In delayed coking, the heated charge (typically residue from atmospheric distillation towers) is transferred to large coke drums which provide the long residence time needed to allow the cracking reactions to proceed to completion. Initially, the heavy feedstock is fed to a furnace which heats the residue to high temperatures (480°-510°C) at low pressures (25-30 psi) and is designed and controlled to prevent premature coking in the heater tubes. The mixture is passed from the heater to one or more coker drums where the hot material is held approximately 24 hours (delayed) at pressures of 25-75 psi, until it cracks into lighter products. Vapours from the drums are returned to a fractionator where gas, naphtha and gas oils are separated out. The heavier hydrocarbons produced in the fractionator are recycled through the furnace.

After the coke reaches a predetermined level in one drum, the flow is diverted to another drum to maintain continuous operation. The full drum is steamed to strip out uncracked hydrocarbons, cooled by water injection, and decoked by mechanical or hydraulic methods. The coke is mechanically removed by an auger rising from the bottom of the drum. Hydraulic decoking consists of fracturing the coke bed with high-pressure water ejected from a rotating cutter.

(b) Continuous Coking : Continuous (contact or fluid) coking is a moving-bed process that operates at temperatures higher than delayed coking. In continuous coking, thermal cracking occurs by using heat transferred from hot, recycled coke particles to feedstock in a radial mixer, called a reactor, at a pressure of 50 psi. Gases and vapours are taken from the reactor, quenched to stop any further reaction, and fractionated. The reacted coke enters a surge drum and is lifted to a feeder and classifier where the larger coke particles are removed as product. The remaining coke is dropped into the preheater for recycling with feedstock. Coking occurs both in the reactor and in the surge drum. The process is automatic, in that, there is a continuous flow of coke and feedstock.

Table 3.3 : Coking process

Feedstock	From	Process	Typical products	To
Residual	Atmospheric & vacuum catalytic cracker	Decomposition	Naphtha, gasoline, column, blending	Distillation
Clarified oil	Catalytic cracker		Coke	Shipping, recycle
Tars	Various units		Gas oil	Catalytic cracking
Wastewater (sour)	Treatment		–	–
Gases	Gas plant		–	–

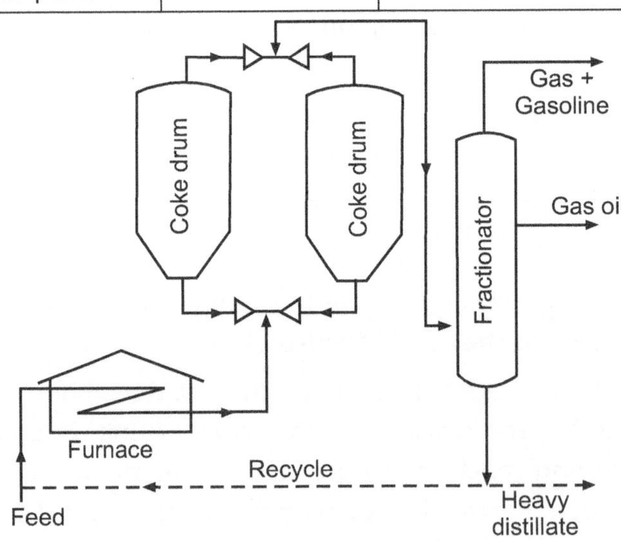

Fig. 3.4 : Delayed coking

3.2.5 Alkylation

Description :

Alkylation combines low-molecular-weight olefins (primarily a mixture of propylene and butylene) with isobutene in the presence of a catalyst, either sulfuric acid or hydrofluoric acid. The product is called *alkylate* and is composed of a mixture of high-octane, branched-chain paraffinic hydrocarbons. Alkylate is a premium blending stock because it has exceptional antiknock properties and clean burning. The octane number of the alkylate depends mainly upon the kind of olefins used, and upon operating conditions.

Sulfuric Acid Alkylation Process : [W-13]

(a) In cascade type sulfuric acid (H_2SO_4) alkylation units, the feedstock (propylene, butylene, amylene, and fresh isobutane) enters the reactor and contacts the concentrated sulfuric acid catalyst (in concentrations of 85% to 95% for good operation and to minimize corrosion). The reactor is divided into zones, with olefins fed through distributors to each zone, and the sulfuric acid and isobutanes flowing over baffles from zone to zone.

(b) The reactor effluent is separated into hydrocarbon and acid phases in a settler, and the acid is returned to the reactor. The hydrocarbon phase is hot-water washed with caustic for pH control before being successively depropanized, deisobutanized and debutanized. The alkylate obtained from the deisobutanizer can then go directly to motor-fuel blending or be rerun to produce aviation-grade blending stock. The isobutane is recycled to the feed.

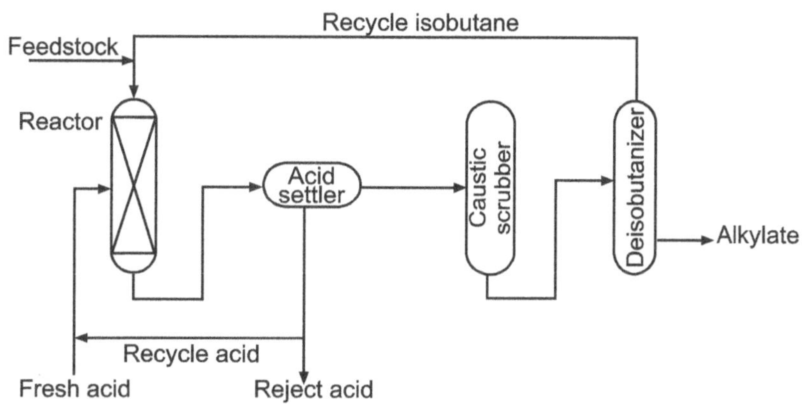

Fig. 3.5 : Sulfuric acid alkylation

Hydrofluoric Acid Alkylation Process :

Phillips and UOP are the two common types of hydrofluoric acid alkylation processes in use. In the Phillips process, olefin and isobutane feedstock are dried and fed to a combination reactor/settler system. Upon leaving the reaction zone, the reactor effluent flows to a settler (separating vessel) where the acid separates from the hydrocarbons. The acid layer at the bottom of the separating vessel is recycled. The top layer of hydrocarbons (hydrocarbon phase), consisting of propane, normal butane, alkylate, and excess (recycle) isobutane, is charged to the main fractionator, the bottom product of which is motor alkylate. The main fractionator overhead, consisting mainly of propane, isobutane and HF, goes to a depropanizer. Propane with trace amount of HF goes to an HF stripper for HF removal and is then catalytically defluorinated, treated, and sent to storage. Isobutane is withdrawn from the main fractionator and recycled to the reactor/settler, and alkylate from the bottom of the main fractionator is sent to product blending.

The UOP Process :

This process uses two reactors with separate settlers. Half of the dried feedstock is charged to the first reactor, alongwith recycle and makeup isobutane. The reactor effluent then goes to its settler, where the acid is recycled and the hydrocarbon charged to the second reactor. The other half of the feedstock also goes to the second reactor, with the settler acid being recycled and the hydrocarbons charged to the main fractionator. Subsequent processing is similar to the Phillips process. Overhead from the main fractionator goes to a depropanizer. Isobutane is recycled to the reaction zone and alkylate is sent to product blending.

Table 3.4 : Alkylation process

Feedstock	From	Process	Typical products	To
Petroleum gas	Distillation or cracking	Unification	High octane gasoline	Blending
Olefins	Catalytic or hydro cracking		n-Butane and propane	Stripper or blender
Isobutane	Isomerization		–	–

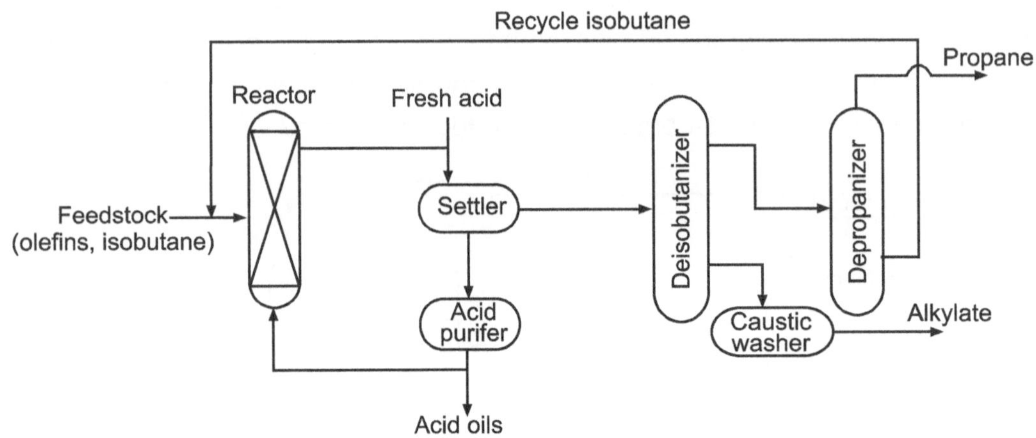

Fig. 3.6 : Hydrogen fluoride alkylation

3.2.6 Polymerization **[W-13; S-16]**

In the petroleum industry, polymerization is used to indicate the production of gasoline components, hence the term "polymer" gasoline. Polymers obtained by the bulk technique are usually pure due to the absence of a solvent. However, in the petrochemical section of a refinery, polymerization, which results in the production of, for instance, polyethylene, is allowed to proceed until materials of the required high molecular weight have been produced.

Methods of Polymerization :

1. Condensation or step growth polymerization.
2. Addition or chain growth polymerization.
   (i) Homogeneous polymerization.
      (a) Bulk polymerization.
      (b) Solution polymerization.
   (ii) Heterogeneous polymerization
      (a) Emulsion polymerization.
      (b) Suspension polymerization.

Description :

(a) Polymerization in the petroleum industry is the process of converting light olefin gases including ethylene, propylene and butylene into hydrocarbons of higher molecular weight and higher octane number that can be used as gasoline blending stocks. Polymerization combines two or more identical olefin molecules to form a single molecule with the same elements in the same proportions as the original molecules. Polymerization may be accomplished thermally or in the presence of a catalyst at lower temperatures.

(b) The olefin feedstock is pretreated to remove sulfur and other undesirable compounds. In the catalytic process, the feedstock is either passed over a solid phosphoric acid catalyst or comes in contact with liquid phosphoric acid, where an exothermic polymeric reaction occurs. This reaction requires cooling water and the injection of cold feedstock into the reactor to control temperatures between 150° to 230°C, at pressures from 200 psi to 1,200 psi. The reaction products leaving the reactor are sent to stabilization and/or fractionator systems to separate saturated and unreacted gases from the polymer gasoline product.

Table 3.5 : Polymerization process

Feedstock	From	Process	Typical products	To
Olefins	Cracking processes	Unification	High octane naphtha	Gasoline blending
			Petrochemical feedstock	Petrochemical industry
			Liquefied petroleum gas	Storage

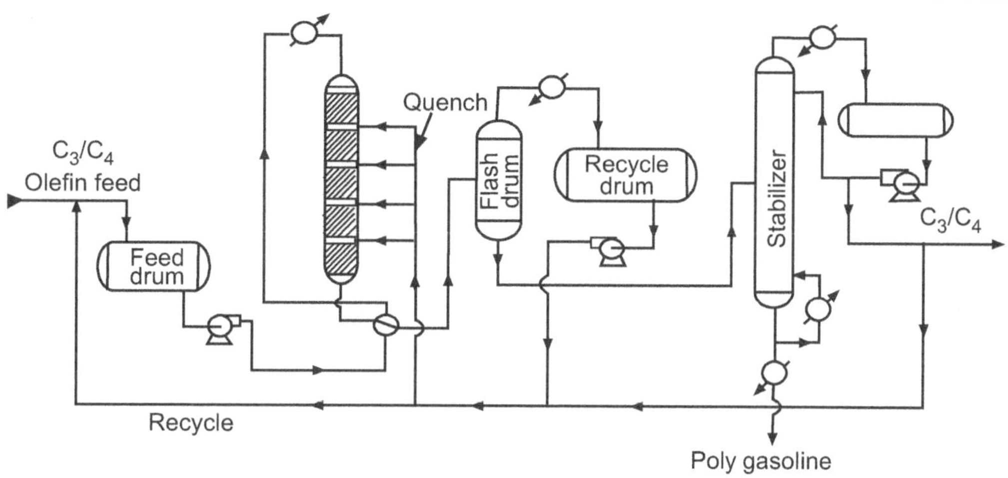

Fig. 3.7 : Polymerization process

3.2.7 Hydrocracking

[S-14, 17, 18; W-17]

Description :

(a) Hydrocracking is a two-stage process combining catalytic cracking and hydrogenation, wherein heavier feedstocks are cracked in the presence of hydrogen to produce more desirable products. The process employs high pressure, high temperature, a catalyst and hydrogen. Hydrocracking is used for feedstocks that are difficult to process by either catalytic cracking or reforming, since these feedstocks are characterized usually by high polycyclic aromatic content and/or high concentrations of the two principal catalyst poisons, sulfur and nitrogen compounds.

(b) The hydrocracking process largely depends on the nature of the feedstock and the relative rates of the two competing reactions, hydrogenation and cracking. Heavy aromatic feedstock is converted into lighter products under a wide range of very high pressures (1,000-2,000 psi) and fairly high temperatures (400°-800°C), in the presence of hydrogen and special catalysts. When the feedstock has a high paraffinic content, the primary function of hydrogen is to prevent the formation of polycyclic aromatic compounds. Another important role of hydrogen in the hydrocracking process is to reduce tar formation and prevent build-up of coke on the catalyst. Hydrogenation also serves to convert sulfur and nitrogen compounds present in the feedstock to hydrogen sulfide and ammonia.

(c) Hydrocracking produces relatively large amounts of isobutane for alkylation feedstock. Hydrocracking also performs isomerization for pour-point control and smoke-point control, both of which are important in high-quality jet fuel.

Hydrocracking Process :

(a) In the first stage, preheated feedstock is mixed with recycled hydrogen and sent to the first-stage reactor, where catalysts convert sulfur and nitrogen compounds to hydrogen sulfide and ammonia. Limited hydrocracking also occurs.

(b) After the hydrocarbon leaves the first stage, it is cooled and liquefied and run through a hydrocarbon separator. The hydrogen is recycled to the feedstock. The liquid is charged to a fractionator. Depending on the products desired (gasoline components, jet fuel and gas oil), the fractionator is run to cut out some portion of the first stage reactor out-turn. Kerosene-range material can be taken as a separate side-draw product or included in the fractionator bottoms with the gas oil.

(c) The fractionator bottoms are again mixed with a hydrogen stream and charged to the second stage. Since this material has already been subjected to some hydrogenation, cracking and reforming in the first stage, the operations of the second stage are more severe (higher temperatures and pressures). Like the out turn of the first stage, the second stage product is separated from the hydrogen and charged to the fractionator.

Table 3.6 : Hydrocracking process

Feedstock	From	Process	Typical products	To
High pour point	Catalytic cracker, atmospheric and vacuum tower	Decomposition, hydrogenation	Kerosene, jet fuel	Blending
Gas oil	Vacuum tower, coker		Gasoline, distillates	Blending
Hydrogen	Reformer		Recycle, reformer gas	Gas plant

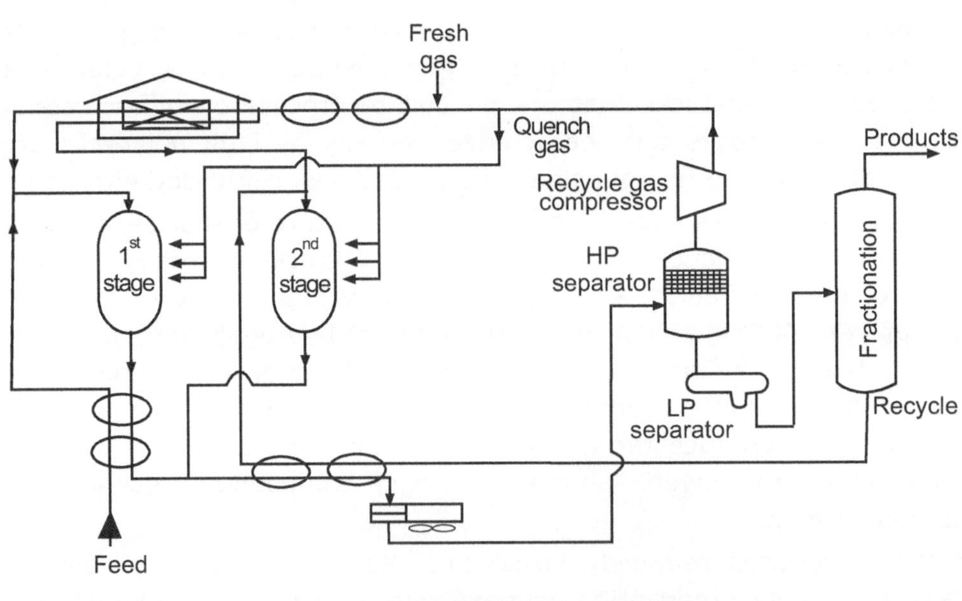

Fig. 3.8 : Two-stage hydrocracking

3.2.8 Isomerization

[W-13; S-17]

Description :

(a) Isomerization is used to alter the arrangement of a molecule without adding or removing anything from the original molecule. Isomerization converts n-butane, n-pentane and n-hexane into their respective isoparaffins of substantially higher octane number. The straight-chain paraffins are converted to their branched-chain counterparts whose component atoms are the same but are arranged in a different geometric structure. Isomerization is important for the conversion of n-butane into isobutane, to provide additional feedstock for alkylation units, and the conversion of normal pentanes and hexanes into higher branched isomers for gasoline blending. Isomerization is similar to catalytic reforming, in that the hydrocarbon molecules are rearranged, but unlike catalytic reforming, isomerization just converts normal paraffins to isoparaffins.

There are two distinct isomerization processes, **Butane (C_4)** and **Pentane/Hexane (C_5/C_6).**

(b) **Butane** isomerization produces feedstock for alkylation. Aluminum chloride catalyst plus hydrogen chloride are universally used for the low-temperature processes. Platinum or another metal catalyst is used for the higher-temperature processes. In a typical low-temperature process, the feed to the isomerization plant is n-butane or mixed butanes mixed with hydrogen (to inhibit olefin formation) and passed to the reactor at 110°-170°C and 200-300 psi. Hydrogen is flashed off in a high-pressure separator and the hydrogen chloride is removed in a stripper column. The resultant butane mixture is sent to a fractionator (deisobutanizer) to separate n-butane from the isobutane product.

(c) **Pentane/hexane** isomerization increases the octane number of the light gasoline components n-pentane and n-hexane, which are found in abundance in straight-run gasoline. In a typical C_5/C_6 isomerization process, dried and desulfurized feedstock is mixed with a small amount of organic chloride and recycled hydrogen, and then heated to reactor temperature. It is then passed over supported-metal catalyst in the first reactor where benzene and olefins are hydrogenated. The feed next goes to the isomerization reactor where the paraffins are catalytically isomerized to isoparaffins. The reactor effluent is then cooled and subsequently separated in the product separator into two streams: a liquid product (isomerate) and a recycle hydrogen-gas stream. The isomerate is washed (caustic and water), acid stripped, and stabilized before going to storage.

Table 3.7 : Isomerization processes

Feedstock	From	Process	Typical products	To
n-Butane n-Pentane n-Hexane	Various Processes	Rearrangement	Isobutane	Alkylation
			Isopentane	Blending
			Isohexane	Blending
			Gas	Gas Plant

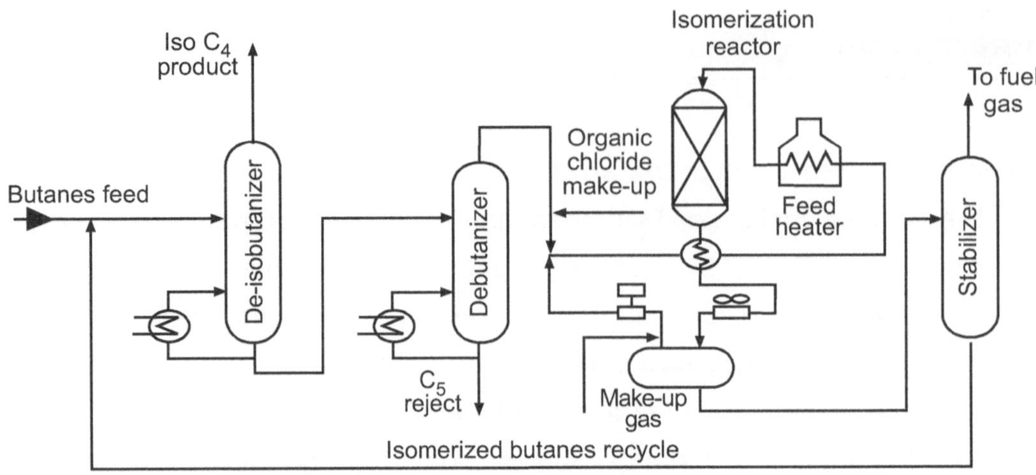

Fig. 3.9 : C_4 Isomerization

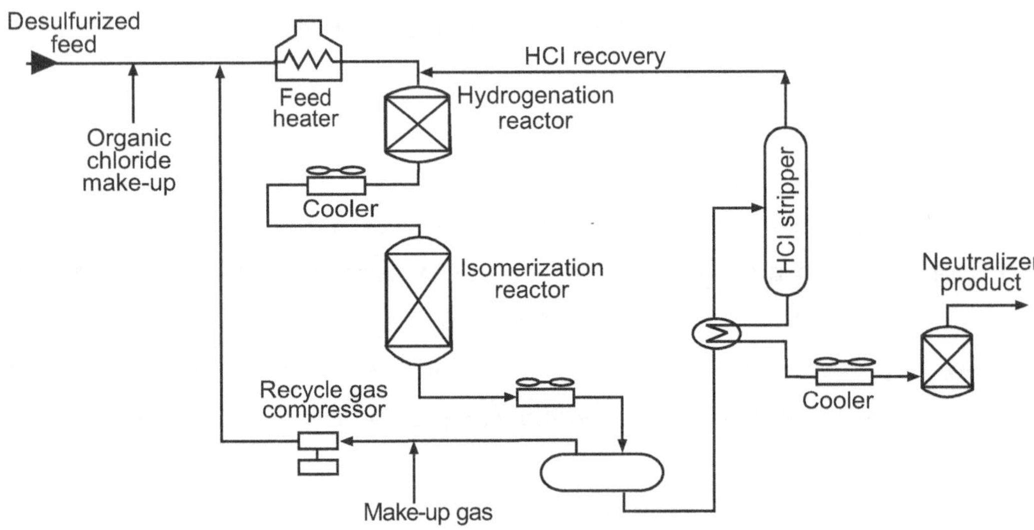

Fig. 3.10 : C_5 and C_6 Isomerization

3.2.9 Catalytic Reforming

[W-13, 17; S-14, 16, 18]

Description :

(a) Catalytic reforming is an important process used to convert low-octane naphthas into high-octane gasoline blending components called *reformates*. Reforming represents the total effect of numerous reactions such as cracking, polymerization, dehydrogenation and isomerization taking place simultaneously. Depending on the properties of the naphtha feedstock (as measured by the paraffin, olefin, naphthene and aromatic content) and catalysts used, reformates can be produced with very high concentrations of toluene, benzene, xylene and other aromatics useful in gasoline blending and petrochemical processing. Hydrogen, a significant by-product, is separated from the reformate for recycling and use in other processes.

(b) A catalytic reformer comprises a reactor section and a product-recovery section. More or less standard is a feed preparation section in which, by combination of hydrotreatment and distillation, the feedstock is prepared to specification. Most processes use platinum as the active catalyst. Sometimes platinum is combined with a second catalyst (bimetallic catalyst) such as rhenium or another noble metal.

(c) There are many different commercial catalytic reforming processes including platforming, powerforming, ultraforming and Thermofor catalytic reforming. In the platforming process, the first step is preparation of the naphtha feed to remove impurities from the naphtha and reduce catalyst degradation. The naphtha feedstock is then mixed with hydrogen, vapourized, and passed through a series of alternating furnace and fixed-bed reactors containing a platinum catalyst. The effluent from the last reactor is cooled and sent to a separator to permit removal of the hydrogen-rich gas stream from the top of the separator for recycling. The liquid product from the bottom of the separator is sent to a fractionator called a *stabilizer* (butanizer). It makes a bottom product called reformate; butanes and lighter go overhead and are sent to the saturated gas plant.

(d) Some catalytic reformers operate at low pressure (50-200 psi), and others operate at high pressures (up to 1,000 psi). Some catalytic reforming systems continuously regenerate the catalyst in other systems. One reactor at a time is taken off-stream for catalyst regeneration, and some facilities regenerate all of the reactors during turnarounds.

Table 3.8 : Catalytic reforming process

Feedstock	From	Process	Typical products	To
Desulfurized naphtha	Coker	Rearrange, dehydrogenate	High octane gasoline	Blending
			Aromatics	Petrochemical
Naphthene-rich fractions	Hydrocracker, hydrodesulfur		Hydrogen	Recycle, hydrotreat etc.
Straight-run naphtha	Atmospheric fractionator		Gas	Gas plant

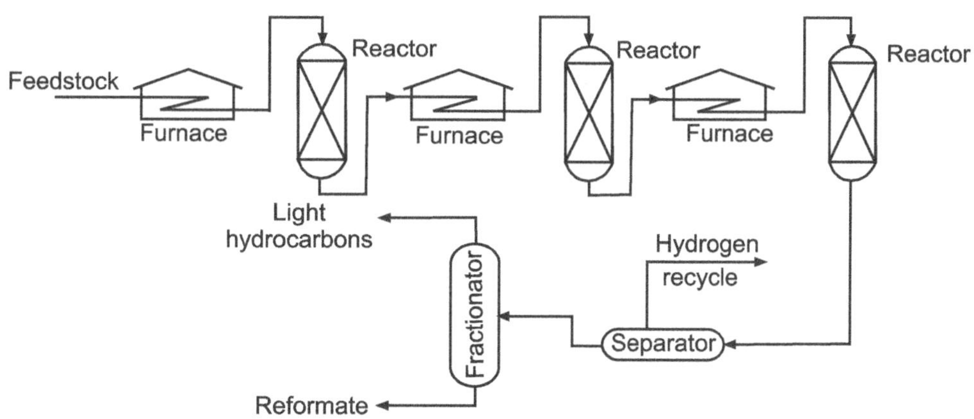

Fig. 3.11 : Reforming process

3.2.10 Catalytic Hydrotreating

Description :

Catalytic hydrotreating is a hydrogenation process used to remove about 90% of contaminants such as nitrogen, sulfur, oxygen and metals from liquid petroleum fractions. These contaminants, if not removed from the petroleum fractions as they travel through the refinery processing units, can have detrimental effects on the equipment, the catalysts, and the quality of the finished product. Typically, hydrotreating is done prior to processes such as catalytic reforming so that the catalyst is not contaminated by untreated feedstock. Hydrotreating is also used prior to catalytic cracking to reduce sulfur and improve product yields, and to upgrade middle-distillate petroleum fractions into finished kerosene, diesel fuel, and heating fuel oils. In addition, hydrotreating converts olefins and aromatics to saturated compounds.

Catalytic Hydrodesulfurization Process :

Hydrotreating for sulfur removal is called hydrodesulfurization. In a typical catalytic hydrodesulfurization unit, the feedstock is deaerated and mixed with hydrogen, preheated in a fired heater (310°-425°C) and then charged under pressure (upto 1,000 psi) through a fixed-bed catalytic reactor. In the reactor, the sulfur and nitrogen compounds in the feedstock are converted into H_2S and NH_3. The reaction products leave the reactor and after cooling to a low temperature enter a liquid/gas separator. The hydrogen-rich gas from the high-pressure separation is recycled to combine with the feedstock, and the low-pressure gas stream rich in H_2S is sent to a gas treating unit where H_2S is removed. The clean gas is then suitable as fuel for the refinery furnaces. The liquid stream is the product from hydrotreating and is normally sent to a stripping column for removal of H_2S and other undesirable components. In cases where steam is used for stripping, the product is sent to a vacuum drier for removal of water. Hydrodesulfurized products are blended or used as catalytic reforming feedstock.

Table 3.9 : Hydrodesulfurization process

Feedstock	From	Process	Typical products	To
Naphthas, distillates sour gas oil, residuals	Atmospheric and vacuum tower, catalytic and thermal cracker	Treating, hydrogenation	Naphtha	Blending
			Hydrogen	Recycle
			Distillates	Blending
			H_2S, ammonia	Sulfur plant, treater
			Gas	Gas plant

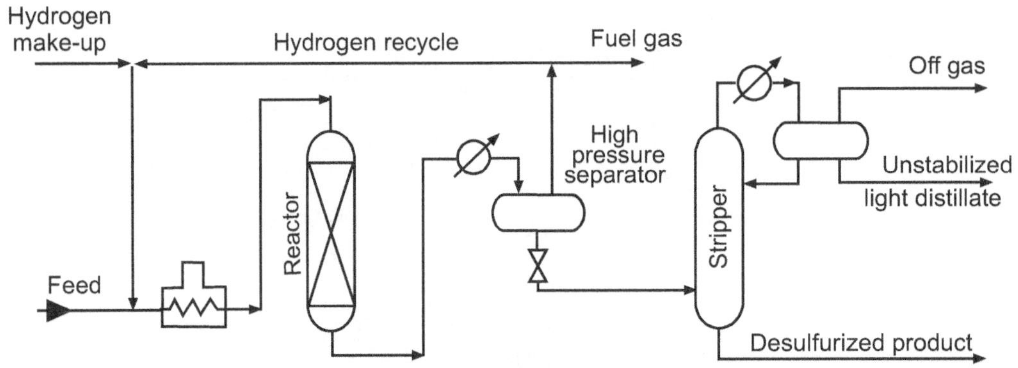

Fig. 3.12 : Distillate hydrodesulfurization

3.2.11 Other Hydrotreating Processes

(a) Hydrotreating processes differ depending upon the feedstock available and catalysts used. Hydrotreating can be used to improve the burning characteristics of distillates such as kerosene. Hydrotreatment of a kerosene fraction can convert aromatics into naphthenes, which are cleaner-burning compounds.

(b) Lube-oil hydrotreating uses catalytic treatment of the oil with hydrogen to improve product quality. The objectives in mild lube hydrotreating include saturation of olefins and improvements in colour, odour and acid nature of the oil. Mild lube hydrotreating also may be used following solvent processing. Operating temperatures are usually below 310°C and operating pressures below 800 psi. Severe lube hydrotreating, at temperatures in the 310°-400°C range and hydrogen pressures upto 3,000 psi, is capable of saturating aromatic rings, along with sulfur and nitrogen removal, to impart specific properties not achieved at mild conditions.

(c) Hydrotreating also can be employed to improve the quality of pyrolysis gasoline (pygas), a by-product from the manufacture of ethylene. Traditionally, the outlet for pygas has been motor gasoline blending, a suitable route in view of its high octane number. However, only small portions can be blended untreated owing to the unacceptable odour, colour and gum-forming tendencies of this material. The quality of pygas, which is high in diolefin content, can be satisfactorily improved by hydrotreating, whereby conversion of diolefins into mono-olefins provides an acceptable product for motor gas blending.

3.2.12 Hydration

[S-17]

Synthetic Ethyl alcohol is generally thought of as the product of fermentation, but most is made by hydration of ethylene. In an older process ethylene dissolves in sulfuric acid to form ethyl sulfate, then hydrolyzes this to form ethanol. There is always some by-product which can be either sold or recirculated.

$$3H_2C{=}CH_2 + 2H_2SO_4 \longrightarrow C_2H_5HSO_4 + (C_2H_5)_2SO_4$$

$$C_2H_5HSO_4 + (C_2H_5)_2SO_4 + H_2O \longrightarrow 3C_2H_5OH + 2H_2SO_4 \text{ (dil. aq.)}$$

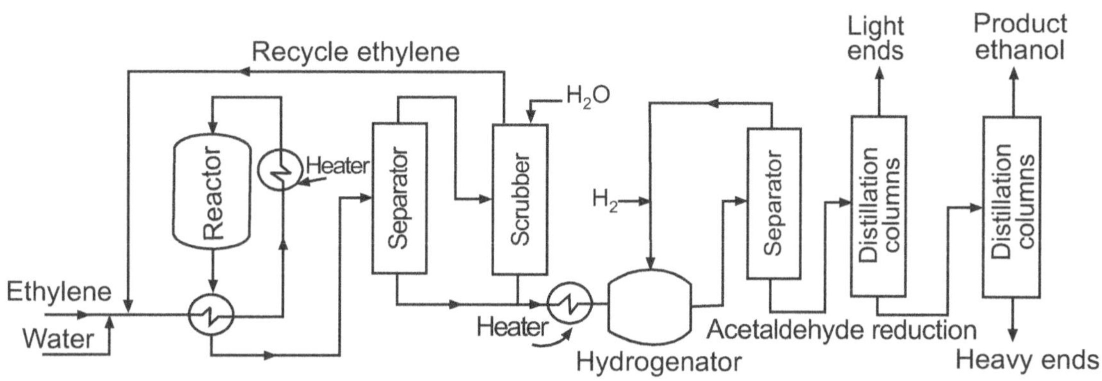

Fig. 3.13: Flow sheet for synthesis of ethyl alcohol from ethylene by direct hydration

The yield is 90% with 5 to 10% as $C_2H_5OC_2H_5$. A direct hydration method is currently favoured for over 85% of production.

$$CH_2{=}CH_2 + H_2O \xrightarrow{H_3PO_4} C_2H_5OH$$

H_3PO_4 catalyst is used at 300°C and 6.8 Mpa which gives a yield of 92%, conversion per pass is 4 to 25 %, depending on the activity of the catalyst used. Ethylene and water are combined with a recycle stream in the ratio ethylene/water = 1/0.6 (mole ratio), a furnace heats the mixture to 300°C, and the gases react over the catalyst of phosphoric acid absorbed on diatomaceous earth. Unreacted reagents are separated and recirculated. Byproduct acetaldehyde is hydrogenated over a catalyst to form more alcohol.

Primarily because of political pressure and government subsidies, the production of fermentation alcohol is growing astronomically. Much ethanol and methanol is finding its way into fuel mixtures, commonly known as *gasohol*. The maximum alcohol content obtainable by simple distillation is 95%, and this will not mix with gasoline. Anhydrous alcohol will mix with gasoline, but the addition of a small amount of moisture will cause it to separate out. It will absorb moisture from air. Despite all these objections, it seems probable that alcohols will ultimately become fuels in substantial volume.

Isopropyl alcohol is a widely used and easily made alcohol, which is the first petrochemical. It is used in making 23% acetone, 10% process solvent, 8% cosmetics, and 6% chemical derivatives. Four process compete: (1) a sulfuric acid process similar to the one described for ethanol hydration, (2) a gas phase hydration using a fixed bed supported phosphoric acid catalyst, (3) a mixed phase reaction using a carbon exchange resin catalyst and (4) a liquid phase hydration in the presence of a dissolved tungsten catalyst. Reaction 2, 3 and 4 are all essentially direct hydration processes.

$$H_3C-CH=CH_2 + H_2O \xrightarrow{\text{catalyst}} CH_3CHOCH_3$$

Per pass conversion vary from (a low) 5% to (a high) of 70% for the gas phase reaction. Reaction 1 is certainly obsolete because of the expense of using sulfuric acid and its pollution problems. The other processes are competitive, non-polluting, and dependent on catalyst life and recycling problems for their economic life.

3.2.13 Esterification **[W-13; S-14]**

In this process, the alcohol and an acid are heated together in presence of sulfuric acid, and the reaction is driven to completion by removing the products as formed and employing an excess of one of the reagents. In the case of ethyl acetate, esterifications take place in a column which takes a ternary azeotrope overhead. Alcohol can be added to the condensed overhead liquid to wash out the alcohol, which is then rectified and returned to the column to react.

Amyl, butyl and isopropyl acetates are all made from acetic acid and the appropriate alcohols. All are useful lacquer solvents and their slow rate of evaporation prevents the surface of the drying lacquer from falling below the dew point, causing condensation on the film and a mottled surface appearance. This phenomenon is known as "blushing". Other esters are used in perfumery and in plasticizers. These include methyl salicylate, methyl anthranilate, diethyl-, dibutyl-, and di 2-ethylhexylphthalates.

Unsaturated vinyl esters for use in polymerization reaction are made by the esterification of olefins. The most important ones are vinyl acetate, vinyl chloride, acrylonitrile and vinyl fluoride. The addition reaction may be carried out in either the liquid, vapour or mixed phases. The choice depends upon the properties of the acid.

Vinyl acetate is manufactured by reacting ethylene with acetic acid in the vapour phase over a supported palladium catalyst.

$$H_2C-CH_2 + CH_3COOH + {}^1/_2\ O_2 \longrightarrow CH_3COOH = CH_2 + H_2O$$

The reaction is conducted in a fixed bed tubular reactor and is highly exothermic. With proper conditions, the only significant byproduct is CO_2 and enough heat is recovered as steam to perform the recovery distillation. Reaction is at 175 to 200°C under a pressure of 475 to 1000 kpa. To prevent polymerization, an inhibitor such as diphenylamine or hydroquinone is added.

Esters of alkyl alcohol, e.g. dialkyl phthalate, are used as bifunctional polymerization monomers. They can be prepared by simple esterification of phthalic-anhydride with alkyl alcohol. Several acrylic esters, such as ethyl or methyl acrylates are also widely used and can be made from acrylic acid and the appropriate alcohol. The esters are more volatile than the corresponding acids.

3.3 PETROCHEMICAL INDUSTRY AND ENVIRONMENT

The Department of Chemicals and Petrochemicals, Government of India (GOI) are the concerned highest authority of the Indian Petrochemical Industry and Environment related issues. The Indian Petrochemical Industry is at par with world standards.

India shares a good portion of the worldwide production of petrochemical business in the world market. Asian, African and even Arab countries buy Indian petrochemical products. The demand for Indian petrochemical products is high, mainly because of its quality and competitive pricing. India's low-cost and high-end petrochemical products, manufacturing expertise, and developing world-class infrastructure are the main leveraging factors for the rise of this industry. India offers petrochemicals at a substantial discount than its western counterparts while delivering the same quality of output.

Advantages of Manufacturing High-class Petrochemical Products in India :

- Friendly Government of India policies.
- Low cost labour.
- Low and world class infrastructure.
- Strong technical education.
- Large number of science and engineering graduates.
- Quality output.
- Highly skilled workforce.
- Usage of innovative process.
- Good client relationships.
- Huge scope for innovation.
- Expansion of existing relationships.
- Huge demand in overseas markets.
- Availability of more technical work force.
- Increased number and quality of training facilities.

Strengths of Indian Petrochemical Industry :

- Large and very fast growing Indian petrochemical market.
- Huge trained talent pool.
- Competitive labour cost.

Weaknesses of Indian Petrochemical Industry :

- Insufficient basic infrastructure for the petrochemical industry.
- High feedstock cost in comparison to middle east countries.
- Prevalence and use of old technology.
- Synthetic fiber industry is unorganized and operates in small clusters.

Opportunities of Indian Petrochemical Industry :

- Huge demand for polymer and synthetic fiber.
- Great opportunity for product development exists.
- Low consumption of polymer in comparison to global consumption rate.

Threats to Indian Petrochemical Industry :

- Stiff competition from other regional players like, China and the Middle East countries.
- Stiff rational pricing pressures.
- Environmental hazards concerns.
- Low market recognition.
- Relocation of manufacturing sites to region with abundance of feedstock.

Notable Points of Indian Petrochemical Industry :

- Represents 2% of world market.
- Does business worth US$ 30 billion.
- Rate of growth of the Indian petrochemical industry is 10%.
- Profit incurred is around 14%.
- Wide variety of products.

- Basic components are petrochemicals, inorganic chemicals and fertilizers.
- The Indian states of Gujarat, Maharashtra, West Bengal and Andhra Pradesh have the largest concentration of chemical and petrochemical units.

In India the following Acts and Rules Govern the Various Aspects of Environment :

- The Water (Prevention and Control of Pollution) Act, 1974.
- The Water (Prevention and Control of Pollution) Cess Act, 1977.
- The Air (Prevention and Control of Pollution) Act, 1981.
- The Environment (Protection) Act, 1986.
- The Batteries (Management and Handling) Rules, 2001.
- The Municipal Solid Wastes (Management and Handling) Rules, 2000.
- The Recycled Plastics Manufacture and Usage Rules, 1999.
- The Rules for the Manufacture, Use, Import and Export and Storage of Hazardous micro-organisms.
- Genetically engineered organisms or cells, 1989.
- The Manufacture, Storage and Import of Hazardous Chemical Rules, 1989.
- The Hazardous Wastes (Management and Handling) Rules, 1989.
- The Bio-Medical Waste (Management and Handling) Rules, 1998.
- Dumping and disposal of fly ash discharged from coal or lignite based thermal power plants on land, 1999.
- Noise Pollution (Regulation and Control) Rules, 2000.
- The Ozone Depleting Substances (Regulation and Control) Rules, 2000.
- The 2-T Oil (Regulation of Supply and Distribution) Order, 1998.
- The Public Liability Insurance Act, 1991.
- The National Environment Appellate Authority Act, 1997.
- The National Environment Tribunal Act, 1995.
- Forest (Conservation) Act, 1980.
- The Biological Diversity Act, 2002.

3.3.1 Sources of Wastewater [S-16]

Relatively large volumes of water are used by the petroleum refining industry. Generally four types of wastewater produced are :

(a) Surface water runoff,

(b) Cooling water,

(c) Process water and

(d) Sanitary wastewater.

Surface water runoff is intermittent and will contain constituents from spills to the surface, leaks in equipment and any materials that may have collected in drains. Runoff surface water also includes water coming from crude and product storage tank roof drains.

A large portion of water used in petroleum refining is used for cooling. Cooling water typically does not come into direct contact with process oil streams and therefore contains less contaminants than process wastewater. Most cooling water is recycled over and over with a bleed or blow down stream to the wastewater treatment unit to control the concentration of contaminants and the solids content in the water. Cooling towers within the recycle loop cool the water using ambient air. (*Storage Tanks and Cooling Towers*) Some cooling water, termed once through, is passed through a process unit once and is then discharged directly without treatment in the wastewater treatment plant. The water used for cooling often contains chemical additives such as chromates, phosphates and antifouling biocides to prevent scaling of pipes and biological growth. (It should be noted, however, that many refineries no longer use chromates in cooling water as anti-fouling agents.) Although cooling water usually does not come into direct contact with oil process streams, it also may contain some oil contamination due to leaks in the process equipment.

Water used in processing operations also accounts for a significant portion of the total wastewater. Process wastewater arises from desalting crude oil, steam stripping operations, pump gland cooling, product fractionator reflux drum drains and boiler blowdown. Because process water often comes into direct contact with oil, it is usually highly contaminated. The quality of the treated waste water and marine environment around the discharge point is monitored by an independent agency all round the year. 70% of the treated effluent is recycled back and used in the Refinery for cooling towers.

Table 3.10 : Typical material output from selected petroleum refining processes

Process	Air Emissions	Process Waste Water	Residual Waste Generated
Crude Oil Desalting	Heater stack gas (CO, SO_X, NO_X, hydrocarbon and particulates), fugitive emissions (hydrocarbons).	Flow - 2.1 Gal/Bbl oil, H_2S, NH_3, phenol, high levels of suspended solids, dissolved solids, high BOD, high temperature.	Crude oil/desalter sludge (iron, rust, clay, sand, water, emulsified oil and wax metals).
Atmospheric Distillation	Heated stack gas (CO, SO_X, NO_X, hydrocarbon and particulates), vents and fugitive emissions (hydrocarbons)	Flow - 26.0 Gal/Bbl oil, H_2S, NH_3, phenol, high levels of suspended solids, dissolved solids, chlorides,	Typically little or no residual waste generated.
Vacuum Distillation	Steam ejector emissions (hydrocarbons), Heater stack gas (CO, SO_X, NO_X, hydrocarbon and particulates), fugitive emissions (hydrocarbons).	Mercaptans, elevated pH.	
Thermal Cracking / Visbreaking	Heater stack gas (CO, SO_X, NO_X, hydrocarbon and particulates), fugitive emissions (hydrocarbons).	Flow - 2.0 Gal/Bbl oil, H_2S, NH_3, suspended solids, high pH, BOD, COD.	Typically little or no residual waste generated.
Coking	Heater stack gas (CO, SO_X, NO_X, hydrocarbon and particulates), fugitive emissions (hydrocarbons) and decoking emissions (hydrocarbon and particulates).	Flow - 1.0 Gal/Bbl oil, high pH, H_2S, NH_3, suspended solids, COD.	Coked dust (carbon particles and hydrocarbons).

... **(Contd.)**

Process	Air Emissions	Process Waste Water	Residual Waste Generated
Catalytic Cracking	Heater stack gas (CO, SO_X, NO_X, hydrocarbon and particulates), fugitive emissions (hydrocarbons) and catalyst regeneration (CO, SO_X, NO_X, hydrocarbon and particulates).	Flow - 15.0 Gal/Bbl High levels of oil, suspended solids, phenols, cyanides, H_2S, NH_3, high pH, BOD, COD.	Spent catalyst (metals from crude oil and hydrocarbon), spent catalyst fines from electrostatic precipitator (aluminum silicate and metals).
Catalytic Hydrocracking	Heater stack gas (CO, SO_X, NO_X, hydrocarbon and particulates), fugitive emissions (hydrocarbons) and catalyst regeneration (CO, SO_X, NO_X, hydrocarbon and catalyst dust).	Flow - 2.0 Gal/Bbl High COD, suspended solids, H_2S, relatively low levels of BOD.	Spent catalyst fines (metals from crude oil and hydrocarbon).
Gas Treatment and Sulfur Recovery	SO_X, NO_X, and H_2S from vent and tail gas emissions.	H_2S, NH_3, amines, stretford solution.	Spent catalyst.
Blending	Fugitive emissions (hydrocarbons).	Little or no waste water generated.	Little or no residual waste generated.
Heat Exchanger Cleaning	Periodic fugitive emissions (hydrocarbons).	Oily wastewater generated.	Heat exchanger sludge (oil, metals and suspended solids).
Storage Tanks	Fugitive emissions (hydrocarbons).	Water drained from tanks contaminated with tank product.	Tank bottom sludge (iron rust, clay, sand, water, emulsified oil and wax, metals).
Blowdown Flare	Combustion products (CO, SO_X, NO_X, hydrocarbon) from flares, fugitive emissions (hydrocarbons).	Little or no waste water generated.	Little or no residual waste generated.

Other Wastes:

Other wastes are generated from many of the refining processes, petroleum handling operations, as well as wastewater treatment. Both hazardous and non-hazardous wastes are generated, treated and disposed. Residual refinery wastes are typically in the form of sludges, spent process catalysts, filter clay and incinerator ash. Treatment of these wastes includes incineration, land treating off-site, land filling on site, land filling off-site, chemical fixation, neutralization, and other treatment methods.

A significant portion of the non-petroleum product outputs of refineries is transported off-site and sold as byproducts. These outputs include sulfur, acetic acid, phosphoric acid and recovered metals. Metals from catalysts and from the crude oil that have deposited on the catalyst during the production often are recovered by third party recovery facilities.

Industrial Effluents :

Wastewater from manufacturing or chemical processes in industries contributes to water pollution. Industrial wastewater usually contains specific and readily identifiable chemical compounds. During the last fifty years, the number of industries in India has grown rapidly. But water pollution is concentrated within a few subsectors, mainly in the form of toxic wastes and organic pollutants. Out of this, a large portion can be traced to the processing of industrial chemicals and to the food products industry. In fact, a number of large and medium-sized industries in the region covered by the Ganga Action Plan do not have adequate effluent treatment facilities. Most of these defaulting industries are sugar mills, distilleries, leather processing industries and thermal power stations. Most major industries have treatment facilities for industrial effluents. But this is not the case with small-scale industries, which cannot afford enormous investments in pollution control equipment as their profit margin is very slender.

Table 3.11 : Different kinds of wastes generated by petroleum refineries

Pollution	Approximate Quantities
Cooling systems	3.5 - 5 m^3 of wastewater generated per ton of crude.
Polluted wastewater	BOD 150 - 250 mg/l. COD 300 - 600 mg/l. Phenol 20 - 200 mg/l. Oil 100 - 300 mg/l (desalter water). Oil 5000 mg/l in tank bottom. Benzene 1 - 100 mg/l. Heavy metals 0.1 - 100 mg/l.
Solid waste and sludge	3 - 5 kg per ton of crude (80% should be considered as hazardous waste because of the heavy metals and toxic organic presence).
VOC emissions	0.5 - 6 kg/ton of crude.
Other emissions	BTX (Benzene, Toluene and Xylene) 0.75 - 6 g/ton of crude. Sulfur oxides 0.2 - 0.6 kg/ton of crude. Nitrogen oxides 0.006 - 0.5 kg/ton of crude.

Effects of Water Pollution :

The effects of water pollution are not only devastating to people but also to animals, fish and birds. Polluted water is unsuitable for drinking, recreation, agriculture and industry. It diminishes the aesthetic quality of lakes and rivers. More seriously, contaminated water destroys aquatic life and reduces its reproductive ability. Eventually, it is hazard to human health. Nobody can escape the effects of water pollution.

The individual and the community can help minimize water pollution. By simple housekeeping and management practices the amount of waste generated can be minimized.

3.3.2 Wastewater Management in the Petrochemical Industry

Petrochemicals can be defined as 'a large group of chemicals derived from petroleum and natural gas and can be used for a variety of chemical purposes'. Petrochemical plants are in the business of developing substances such as hydrogen, carbon monoxide, synthesis gases, chemicals such as ethylene and its derivatives, benzene and toluene to name a few.

Wastewater treatment in petroleum refineries is a complex process, with demanding environmental management challenges, as byproducts can be both volatile and toxic. Petrochemical wastewater often requires a combination of treatment methods to remove oil and other contaminants before discharge. Issues such as groundwater contamination, aromatics, oil, grease and organic removal and VOC control have to be addressed in order to comply with environmental regulations and maintain a good customer image.

Table 3.12 : Wastewater treatment of Petrochemical Industry, (2012-18)

Year	Equipment revenues ($Million)	Growth rate (%)
2012	89.5	-
2013	95.0	6.1
2014	101.0	6.3
2015	107.5	6.4
2016	114.5	6.5
2017	122.5	7.0
2018	131.5	7.3
Compound Annual Growth Rate (CAGR) 2012-18		6.6

A typical wastewater system may include neutralization, coagulation/flocculation, floatation/ sedimentation/filtration, clarification and biodegradation (e.g., trickling filter, anaerobic treatment and aerated lagoon, rotating biological contactor and activated sludge). A final polishing step using filtration, ozonation, activated carbon, or chemical treatment may also be required.

An increased understanding of requirements of the petrochemical end-user is gaining importance as industry needs change due to stricter regulation and environmental laws. Its important equipment manufacturers keep abreast of these requirements to capitalize on future opportunities. End-users are forced to become more aware of what types of waste they are creating and discharging, and how they can adjust their current treatment equipment to meet new demands.

Pollution Prevention Opportunities :

The best way to reduce pollution is to prevent it in the first place. Some companies have creatively implemented pollution prevention techniques that improve efficiency and increase profits while at the same time minimize environmental impacts. This can be done in many ways such as reducing material inputs, re-engineering processes to reuse by-products, improving management practices, and employing substitution of toxic chemicals. Some smaller facilities are able to actually get below regulatory thresholds just by reducing pollutant releases through aggressive pollution prevention policies.

In order to encourage these approaches, this section provides both general and company-specific descriptions of some pollution prevention advances that have been implemented within the petroleum refining industry. While the list is not exhaustive, it does provide core information that can be used as the starting point for facilities interested in beginning their own pollution prevention projects. When possible, this section provides information from real activities that can be, or are being, implemented by this sector - including a discussion of associated costs, time frames, and expected rates of return. This section provides summary information from activities that may be, or are being implemented by this sector. When possible, information is provided that gives the context in which the technique can be effectively used. Please note that the activities described in this section do not necessarily apply to all facilities that fall within this sector. Facility-specific conditions must be carefully considered when pollution prevention options are evaluated, and the full impacts of the change must examine how each option affects air, land and water pollutant releases.

3.4 WASTEWATER TREATMENT [S-14, 16, 17, 18; W-13, 17]

The current wastewater treatment method includes collection allowing different streams to be collected separately, routing and treatment. Dependent of the degree of the (required) treatment the following steps are distinguished:

(a) Primary treatment (API/CPI/PPI separator, sour water strippers, buffer tanks).

(b) Secondary treatment (coagulation, flocculation-flotation, biological treatment).

(c) Tertiary treatment (sand-filtration, membrane-filtration, chemical oxidation).

(d) Pre-treatment.

Pollution Characteristics of Petrochemical Waste and its Treatment:

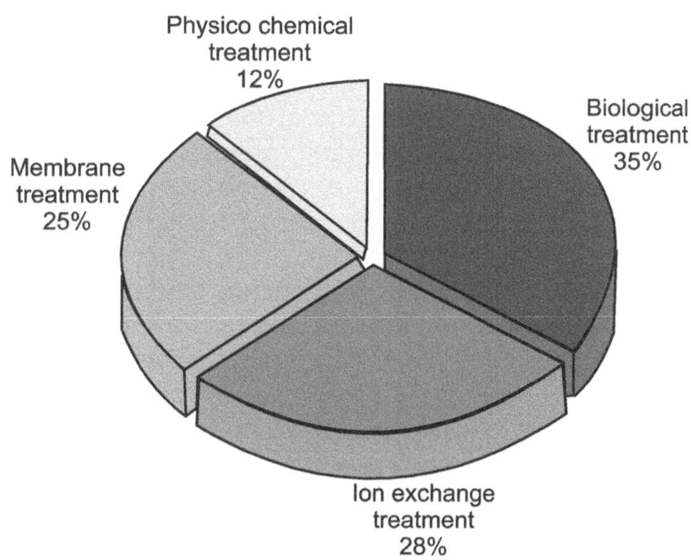

Fig. 3.14 : Breakdown by wastewater treatment equipment in the petrochemical industry (U.S.)

Above Fig. 3.14 gives a split of the different equipment categories for petrochemical wastewater treatment. It is interesting to note advanced biological treatment methods still form a sizable portion of wastewater treatment equipment for petrochemical applications. This is because there is a growing preference in the industry for non-chemical treatment processes. This upward trend provides growth opportunities for suppliers in advanced biological systems such as membrane bioreactors in the petrochemical wastewater segment.

Table 3.13 : Characteristics and its treatment

Pollution Characteristics	Conventional Treatment
1. Oil 2. High BOD and COD 3. High total solids 4. Hydrocarbons and their products	Chemical treatment and biological treatment

Wastewater treatment is used for process water, runoff water and sewerage water prior to discharge or recycling. Wastewater typically contains hydrocarbons, dissolved materials, suspended solids, phenols, ammonia, sulfides and other compounds. Wastewater includes condensed steam, stripping water, spent caustic solutions, cooling tower and boiler blowdown, wash water, alkaline and acid waste, neutralization water, and other process-associated water. Petroleum refineries typically utilize **primary** and **secondary** wastewater treatment.

(a) Primary Treatment :

Primary treatment is the separation of oil, water, hydrocarbons and solids from wastewater in two stages. During the first stage, an API separator, a corrugated plate interceptor, or other separator design is used. Wastewater moves very slowly through the separator allowing free oil to float to the surface and be skimmed off, and solids to settle to the bottom and be scraped off to a sludge collecting hopper. The second stage utilizes physical or chemical methods to separate emulsified oils from the wastewater. Physical methods may include the use of a series of settling ponds with a long retention time, or the use of dissolved air flotation (DAF). In DAF, air is bubbled through the wastewater, and both oil and suspended solids are skimmed off the top. Chemicals, such as ferric hydroxide or aluminum hydroxide, can be used to coagulate impurities into a froth or sludge which can be more easily skimmed off the top. Some wastes associated with the primary treatment of wastewater at petroleum refineries may be considered hazardous and include: API separator sludge, primary treatment sludge, sludges from other gravitational separation techniques, float from DAF units, and wastes from settling ponds.

(b) Secondary Treatment :

After primary treatment, the wastewater undergoes a secondary treatment process. In secondary treatment, dissolved oil and other organic pollutants may be consumed biologically by microorganisms. Biological treatment may require the addition of oxygen through a number of different techniques, including activated sludge units, trickling filters, and rotating biological contactors. Secondary treatment generates bio-mass waste which is typically treated anaerobically, and then dewatered. These processes biologically degrade and oxidize soluble organic matter by the use of activated sludge, unaerated or aerated lagoons, trickling filter methods, or anaerobic treatments. Materials with high adsorption characteristics are used in fixed-bed filters or added to the wastewater to form slurry which is removed by sedimentation or filtration. Additional treatment methods are used to remove oils and chemicals from wastewater.

Some refineries employ an additional stage of wastewater treatment called *polishing* to meet discharge limits. The polishing step can involve the use of activated carbon, anthracite coal, or sand to filter out any remaining impurities, such as biomass, silt, trace metals and other inorganic chemicals, as well as any remaining organic chemicals.

Certain refinery wastewater streams are treated separately, prior to the wastewater treatment plant, to remove contaminants that would not easily be treated after mixing with other wastewater. One such waste stream is the sour water drained from distillation reflux drums. Sour water contains dissolve hydrogen sulfide and other organic sulfur compounds and ammonia which are stripped in a tower with gas or steam before being discharged to the wastewater treatment plant.

Wastewater treatment plants are a significant source of refinery air emissions and solid wastes. Air releases arise from fugitive emissions from the numerous tanks, ponds and sewer system drains. Solid wastes are generated in the form of sludges from a number of the treatment units.

(c) Tertiary Treatment Operations :

Tertiary treatments remove specific pollutants to meet regulatory discharge requirements. These treatments include chlorination, ozonation, ion exchange, reverse osmosis, activated carbon adsorption etc. Compressed oxygen is diffused into wastewater streams to oxidize certain chemicals or to satisfy regulatory oxygen-content requirements. Wastewater that is to be recycled may require cooling to remove heat and/or oxidation by spraying or air stripping to remove any remaining phenols, nitrates and ammonia.

(d) Pre-Treatment Process :

Pre-treatment of used oil involves removing any water within the oil, known as *dewatering*. One way of doing this is by placing it in large settling tanks, which separates the oil and water.

Water is found in used oil as free water or bound water, for example in emulsions. The term dewatering is usually taken to mean the removal of free water. Where water has been emulsified with oil, the emulsion has to be "broken" with a demulsifier before the water can be separated from the oil.

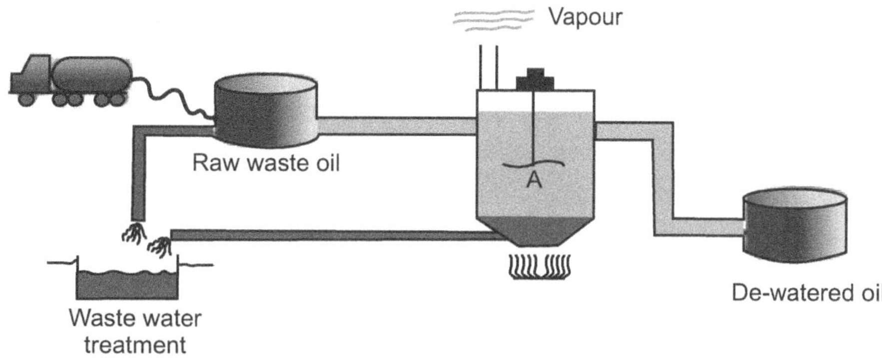

Fig. 3.15 : Pre-treatment process

Dewatering is a simple process relying on the separation of aqueous and oil phases over time under the influence of gravity. The used oil is allowed to stand in a tank (raw waste oil) and free water drops to the bottom where it can be drained, treated (waste water treatment) and discharged appropriately to sewer or stormwater depending on quality and local regulations.

Heating and stirring the used oil in a tank (A) and driving off the water through evaporation can speed up the dewatering process. The "dried" or dehydrated oil is then suitable for further processing or for use as a burner fuel.

Further recycling steps include :

- **Filtering and demineralisation** of the oil, to remove any solids, inorganic material and certain additives present in the oil, producing a cleaner, burner fuel or feed oil for further refining;
- **Propane de-asphalting** to remove the heavier bituminous fractions, producing re-refined base oil; and
- **Distillation** to produce re-refined base oil suitable for use as a lubricant, hydraulic or transformer oil. This process is very similar to the process undergone by virgin oil.

3.4.1 Recycling Steps in Wastewater Treatment [W-13]

The steps are explained in detail below.

- **Filtering and Demineralisation :**

The purpose of filtering and demineralisation is to remove inorganic materials and certain additives from used oil to produce a cleaner burner fuel or feed for re-refining.

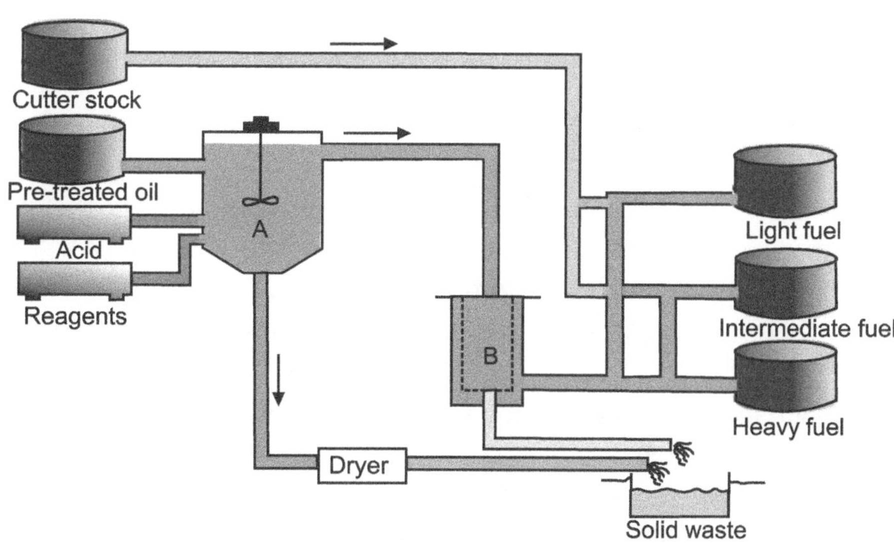

Fig. 3.16 : Filtering and demineralisation process

Used oil feedstock is transferred to a reaction tank (A) and mixed with a small quantity of sulphuric acid and heated to about 60°C. A chemical surface-active reagent, called a *surfactant*, is added to the reactor (A) and after stirring the mixture is allowed to stand. This allows the mixture to separate into two "phases" - i.e. oil and water-based or aqueous.

The reagent causes the contaminants to accumulate in the aqueous phase, which settles to the bottom of the tank (A) and is drained off as slurry. This phase contains acid, used oil contaminants, including metals and some of the oil additives. The water is dried off; leaving a solid waste that must be disposed off. The demineralised oil is filtered (B) to remove suspended fine particles (to solid waste) and runoff to storage (C) as a clean burner fuel. It can be further diluted or "cut" with a lighter petroleum product (called cutter stock) to produce a range of intermediate to light fuel oils depending on the fuel viscosity requirements of the burner.

- **Propane De-asphalting Process :** **[W-13]**

The Propane De-asphalting (PDA) process is an important pre-treatment step in the re-refining process producing de-asphalted lube-oil, which becomes a feedstock for the next step in a re-refining facility. The other output (which is also an input) is propane, which is recovered from both streams and re-used within the process. The PDA process relies on the greater solubility of the paraffinic and naphthenic (i.e. essentially the base oil) components versus the contaminated waste material in a stream of propane. The separation of the lubricating oil fraction from used oil is a continuous process and is conducted at ambient temperature while processing used oil.

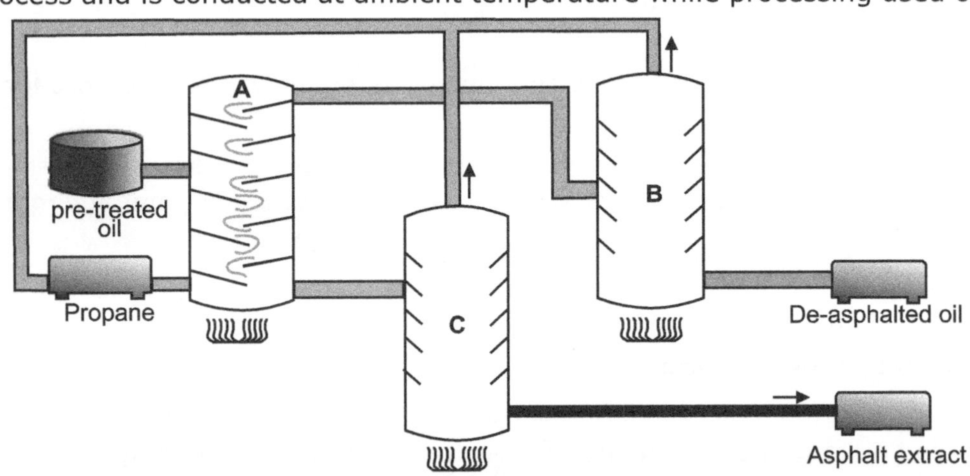

Fig. 3.17 : Propane de-asphalting process

Process Description :

(i) The used oil is pumped into the middle of the extraction column (A). Liquid propane is charged to the bottom of the column (A).

(ii) The oil being heavier than propane flows down the column (A); the propane rises in a counter-flow thus mixing the input streams within the column (A). The rising propane dissolves the more soluble lube oil components, which are carried out to the top of the column (A) with the propane, and the propane insoluble material is removed from the bottom of the column (A).

(iii) Propane is vapourised from both "stripper" units (B) and (C), then condensed and returned to the propane storage tank.

(iv) The de-asphalted lube-oil component is feed for the next processing stage. The residuum (waste) component is mixed with bottoms from the vacuum distillation tower to produce an asphaltic material.

Distillation (or Fractionation) :

It is the physical separation of components of lubricating oil by boiling range. Depending on the type of distillation, the boiling ranges can produce gases and gasolines at the lower boiling points with heavy lubricating oils being distilled at higher boiling points. Distillation is the core process for a facility capable of producing re-refined base-oils to virgin base-oil quality. There are two types of distillation.

(a) Atmospheric Distillation.

(b) Vacuum Distillation.

(a) Atmospheric distillation : It is generally (but not always) considered a pre-treatment step for vacuum distillation and does not require de-watered feedstock. Atmospheric distillation is carried out at normal atmospheric pressure and with temperatures up to 300°C. Prior to the atmospheric distillation process, the feedstock can have undergone PDA treatment, but this is not an absolute pre-requisite.

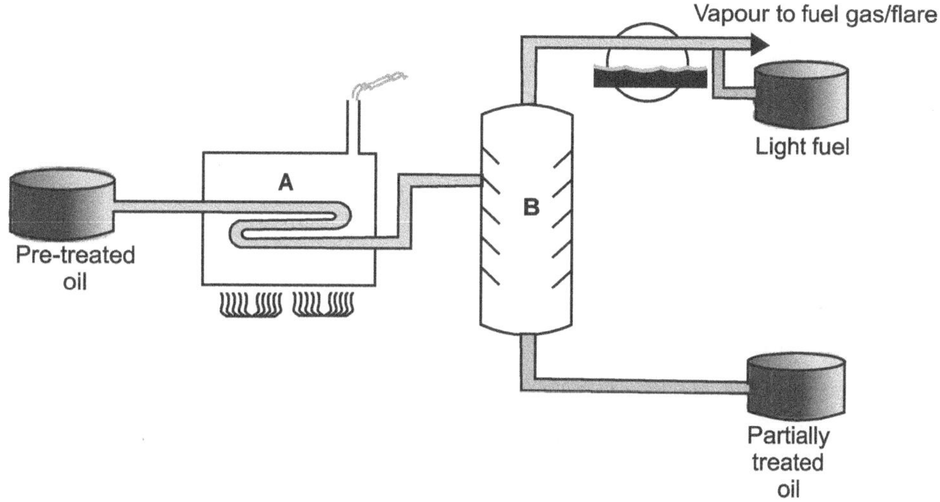

Fig. 3.18 : Atmospheric distillation process

Process Description :

(i) Atmospheric distillation is a relatively simple process separating lower boiling point liquids at ambient pressure. Used oil is heated (A) and charged to a distillation tower (B).

(ii) Lower boiling point hydrocarbons present in the used oil (e.g. gases, petrol and solvents) and water are collected at the top of the tower (B).

(iii) This process is only suitable for temperatures up to 300°C, as temperatures above this can lead to "thermal cracking" of the larger molecule (higher boiling point) hydrocarbons.

(iv) After atmospheric distillation, the oil usually undergoes vacuum distillation. However, it is generally accepted that water and lower boiling point hydrocarbon components be removed prior to vacuum distillation.

(b) Vacuum distillation : It is considered as the key process used in oil re-refining. If atmospheric distillation is utilised, the oil from the atmospheric distillation column is the feedstock for the vacuum distillation column.

In vacuum distillation, the feedstock can be separated into products of similar boiling range to better control the physical properties of the lube base stock "distillate cuts" that will be produced from the vacuum tower products.

The major properties that are controlled by vacuum distillation are viscosity, flash point and carbon residue.

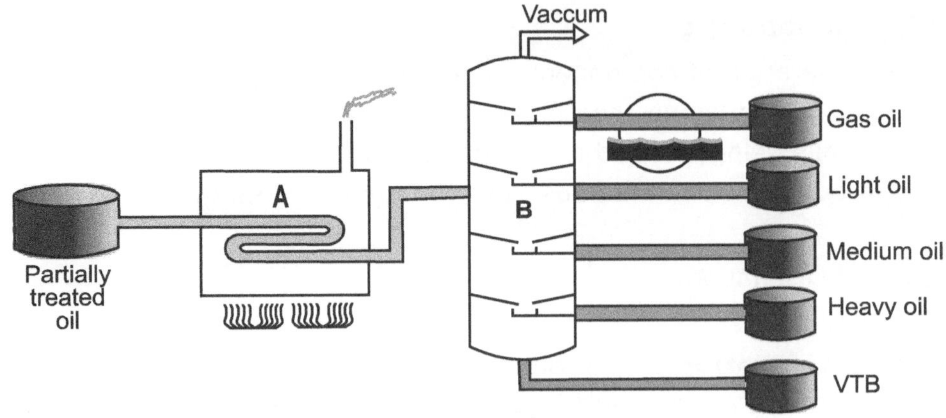

Fig. 3.19 : Vacuum distillation process

Process Description :

(i) The used oil feedstock (usually from the atmospheric distillation unit) is heated in a furnace (A) and flows as a mixture of liquid and vapour to the heated vacuum distillation column (B) where the vapour portion begins to rise and the liquid falls. Steam can be added to assist vapourisation.

(ii) A vacuum is maintained in the column (2-10 mm Hg) by a vacuum system connected to the top of the tower (B). By reducing the pressure, materials are normally boiled upto 540°C.

(iii) At various points in the column (B), special trays, called *draw trays*, are installed which permit the removal of the liquid from the column.

(iv) Some of the material does not boil even under this vacuum. This remains in the vacuum tower and is run out as the vacuum tower bottoms (VTB's). This material contains the heaviest molecules, including some lube oil additives and carryover contaminants not removed in the PDA process.

Important Points

- Unit Processes in Refineries :

  (i) Hydrogenation : It is the addition of H_2 to unsaturated organic compounds such as alkenes to give alkanes and aldehydes to give alcohols.

  $$RCH{=}CH_2 + H_2 \longrightarrow RCH_2CH_3 \text{ (R = alkyl, aryl)}$$

  (ii) Catalytic cracking : It breaks complex hydrocarbons into simpler molecules in order to increase the quality and quantity of lighter, more desirable products in presence of acid catalysts such as amorphous silica-alumina.

  (iii) Fluid catalytic cracking : The fluid cracker consists of a catalyst section and a fractionating section that operate together as an integrated processing unit.

  (iv) Thermal cracking : It breaks or cracks the heavier, higher boiling-point petroleum fractions into more valuable products such as gasoline, fuel oil, and gas oils. Following are applications of thermal cracking: visbreaking, steam cracking, and coking.

  (v) Alkylation : It combines low-molecular-weight olefins. The product is called alkylate and is composed of a mixture of high-octane, branched-chain paraffinic hydrocarbons. This process is classified into Sulfuric acid and Hydrofluoric acid alkylation.

  (vi) Polymerization : It converts light olefin gases including ethylene, propylene, and butylene into hydrocarbons of higher molecular weight and higher octane number that can be used as gasoline blending stocks.

(vii) **Hydrocracking :** It is a two-stage process combining catalytic cracking. It produces relatively large amounts of isobutane for alkylation feedstock.

(viii) **Isomerization :** It is used to alter the arrangement of a molecule without adding or removing anything from the original molecule. Isomerization converts n-butane, n-pentane and n-hexane into their respective isoparaffins. There are two distinct isomerization processes, Butane (C_4) and Pentane/Hexane (C_5/C_6).

(ix) **Catalytic reforming :** It is an important process used to convert low-octane naphthas into high-octane gasoline blending components called reformates.

(x) **Other Hydrotreating Processes :** It employed to improve the quality of pyrolysis gasoline (pygas), a by-product from the manufacture of ethylene. Hydrotreating can be used to improve the burning characteristics of distillates such as kerosene. Hydrotreatment of a kerosene fraction can convert aromatics into naphthenes, which are cleaner-burning compounds.

(xi) **Hydration :** A direct hydration method is currently favoured for over 85% of production. In this process, addition of water molecule take place. By addition of H_3PO_4 catalyst, we can increase the yield of product.

(xii) **Esterification :** Esters of alkyl alcohol, e.g. dialkyl phthalate, are used as bifunctional polymerization monomers. They can be prepared by simple esterification of phthalic-anhydride with alkyl alcohol. Unsaturated vinyl esters for use in polymerization reaction are made by the esterification of olefins. The most important ones are vinyl acetate, vinyl chloride, acrylonitrile and vinyl fluoride.

- **Petrochemical Industry And Environment :** India shares a good portion of the worldwide production of petrochemical business in the world market.
- **Sources of Wastewater :** Main sources of wastewater are :

  (a) Surface water runoff,

  (b) Cooling water,

  (c) Process water and

  (d) Sanitary wastewater.
- **Wastewater Treatment :**

  (a) Primary treatment : It is the separation of oil, water, hydrocarbons and solids from wastewater in two stages. It is a physical or chemical method to separate emulsified oils from the wastewater.

  (b) Secondary treatment : It generates bio-mass waste which is typically treated anaerobically, and then dewatered. These processes biologically degrade and oxidize soluble organic matter by the use of activated sludge, unaerated or aerated lagoons, trickling filter methods, or anaerobic treatments.

  (c) Tertiary treatments : It removes specific pollutants to meet regulatory discharge requirements. These treatments include chlorination, ozonation, ion exchange, reverse osmosis, activated carbon adsorption etc.

  (d) Pre-treatment (Dewatering) process : Dewatering means removing of water within the oil. Dewatering is a simple process relying on the separation of aqueous and oil phases over time under the influence of gravity.

It involves the following steps :

- **Filtering and Demineralisation :** It is to remove inorganic materials and certain additives from used oil to produce a clean burner fuel or feed for re-refining by adding a sulfuric acid and heated to 60°C.
- **Propane De-asphalting (PDA) process :** It is a re-refining process producing de-asphalted lube-oil, which relies on the greater solubility of the paraffinic and naphthenic.
- **Distillation :** In this process, components of lubricating oil are separated by boiling point difference. There are two types : (a) Atmospheric, (b) Vacuum.

Practice Questions

1. Define Alkylate and give the parameters of it.
2. Explain Fluid Catalytic Cracking process.
3. Explain Filtering and demineralization, and Propane De-asphalting process.

MSBTE Questions and Answers (As Per G-Scheme)

Summer 2016

1. Describe any one alkylation process with reactions. **(4 M)**

Ans. Refer Section 3.2.5.

2. What are the characteristics of waste water produced in petrochemical plant? **(4 M)**

Ans. Refer Section 3.3.1.

3. Give the names of two types of cracking processes. State the basic difference between them. **(4 M)**

Ans. Refer Section 3.2.2 and 3.2.3.

4. What is polymerization? State the three methods of polymerization. **(4 M)**

Ans. Refer Section 3.2.6.

5. Explain steps in waste water treatment in petrochemical industry. **(4 M)**

Ans. Refer Section 3.4.

6. Explain reforming with neat flow sheet. **(4 M)**

Ans. Refer Section 3.2.9.

7. Explain any one coking process. **(4 M)**

Ans. Refer Section 3.3.4.

8. Explain thermal cracking process. **(4 M)**

Ans. Refer Section 3.3.4.

9. What is Isomerization ? Explain with flow sheet. **(4 M)**

Ans. Refer Section 3.2.8.

Summer 2017

1. Draw well labeled flow diagram of polymerization process. **(4 M)**

Ans. Refer Section 3.2.6.

2. Describe waste water treatment in petroleum and petrochemical industries, in brief. **(4 M)**

Ans. Refer Section 3.4.

3. Draw well labelled flow diagram of isomerization process. **(4 M)**

Ans. Refer Section 3.2.8.

4. Describe alkylation process with well labelled flow diagram. **(4 M)**

Ans. Refer Section 3.2.5.

5. Describe hydrocracking process with well labelled flow diagram. **(4 M)**

Ans. Refer Section 3.2.7.

6. Define following term with suitable example **(4 M)**
   (i) Thermal cracking
   (ii) Catalytic cracking

Ans. Refer Sections 3.3.4 and 3.2.2 respectively.

7. Write any four chemical reactions involved in reforming process with standard conditions. **(4 M)**

Ans. Refer Section 3.2.9.

8. Describe hydration process with well labelled flow diagram. **(4 M)**

Ans. Refer Section 3.2.12.

9. Describe hydrogenation process with well labelled flow diagram. **(4 M)**

Ans. Refer Section 3.2.1.

Winter 2017

1. Describe any one alkylation process with flow sheet. **(6 M)**

Ans. Refer Section 3.2.5.

2. Explain any one method of hazardous waste treatment. **(4 M)**

Ans. Refer Section 3.4.

3. What is the difference between thermal cracking and catalytic cracking? What are the main advantages of catalytic cracking? **(4 M)**

Ans. Refer Section 3.2.3 and 3.3.4.

4. What is polymerization? State three methods of polymerization. **(4 M)**

Ans. Refer Section 3.2.6.

5. Explain moving bed catalytic cracking process with flow sheet. **(6 M)**

Ans. Refer Section 3.2.3.1.

6. What is catalytic cracking process-fluidized bed type? Explain with flow sheet. **(8 M)**

Ans. Refer Section 3.2.3.

7. Draw the flow sheet for high pressure platinum catalyst reforming process. Explain the process. **(8 M)**

Ans. Refer Section 3.2.9.

8. Explain the process of hydrocracking with flow sheet and reactions involved in it. **(8 M)**

Ans. Refer Section 3.2.7.

Summer 2018

1. Explain any one method of hazardous waste treatment in petrochemical industry. **(4 M)**

Ans. Refer Section 3.4.

2. Differentiate between catalytic cracking and thermal cracking. **(4 M)**

Ans. Refer Section 3.2.3.

3. Explain principle of reforming with reactions involved in it. **(4 M)**

Ans. Refer Section 3.2.9.

4. Explain any one isomerization process with neat flow diagram. **(8 M)**

Ans. Refer Section 3.2.8.

5. Explain : **(8 M)**

   (i) Hydrogenation.

   (ii) Hydrocracking with neat flow sheet.

Ans. Refer Section 3.2.1 and 3.2.7.

6. Explain the H_2SO_4 alkylation process. **(4 M)**

Ans. Refer Section 3.2.5.

7. What is visbreaking ? Describe it with neat diagram. **(4 M)**

Ans. Refer Section 3.3.4.

8. Explain : Esterification process. **(4 M)**

Ans. Refer Section 3.2.13.

✍✍✍

Chapter **4...**

PETROCHEMICALS (FROM C_1 TO C_3)

Weightage of Marks = 16, Teaching Hours = 10

Syllabus

4.1 Petrochemical industries in India, List of chemicals from C_1, Manufacturing of Methanol and Formaldehyde.

4.2 Esterification process.

4.3 List of chemicals from C_2, Manufacturing of Ethanol, Vinyl chloride, Ethylene oxide and Styrene from C_2.

4.4 List of chemicals from C_3, Manufacturing of Propylene Oxide and Acetaldehyde.

About this Chapter

At the end of this chapter, students will be able to:

- Explain the manufacturing of the specified petrochemical of C_1/C_2 or C_3 with process flow diagram.
- State the reaction of the specified petrochemical of C_1/C_2 or C_3 for its production.
- Draw the flow sheet for manufacturing of specified C_1/C_2 or C_3 petrochemical.

4.1 PETROCHEMICAL INDUSTRIES IN INDIA

The petrochemical industry in India has been one of the fastest growing industries in the country. Since the beginning, the Indian petrochemical industry has shown an enviable growth rate. This industry also contributes largely to the economy of the country and the growth and development of manufacturing industry as well. It provides the foundation for manufacturing industries like construction, packaging, pharmaceuticals, agriculture, textiles etc.

The Indian petrochemical industry is a highly concentrated one and is oligopolistic in nature. Even till a few days back, only four major companies viz. Reliance Industries Ltd (RIL), Indian Petrochemicals Corporation Ltd. (IPCL), Gas Authority of India Ltd. (GAIL) and Haldia Petrochemicals Ltd. (HPL) used to dominate the industry to a large extent. The recent amalgamation of IPCL with RIL has made the industry more concentrated further, as they jointly account for over 70% of country's total petrochemical capacity. However, the scene is a bit different for the downstream petrochemical sector, which is highly fragmented in nature with over 40 companies existing in the market.

Growth of Indian Petrochemical Industries:

The petrochemical industry in India came into existence during 1970s. The 1980s and 1990s saw some rapid growths for Indian petrochemical industry. The biggest reason for this growth was the high demand for petrochemicals in India, which grew at an annual rate of 13 to 14% since the late 90s. It also called for rapid expansion of capacity. The BMI forecast of average annual growth in India from 2007-2011 is 14 to 16 %. However, the industry suffered setbacks during 2008 due to surge in the price of crude oil. It will be tough for Indian petrochemical industry to plug the deficit of 5mn TPA of ethylene and 4mn TPA of polymer by 2012 (according to the predictions of the government).

Presently India has three gas-based and three naphtha-based cracker complexes with a combined annual capacity of 2.9 MMT of ethylene. Besides this, there are also 4 aromatic complexes with a capacity of 2.9 MMT of Xylenes.

The production of 5.06 MMT polymers during Financial Year 2009 accounted for around 62% of the total production of key petrochemicals. It also achieved 88.5% capacity utilization. The industry also produced 2.52 MMT of synthetic fibers during Financial Year 2009 with a 73% of capacity utilization.

Key Issues and Challenges:

The petrochemical industry in India is faced with numerous hurdles that prevent it from surpassing other Asian countries.

- India's production of ethylene is far lesser than China. According to experts, the condition is going to remain so for at least the next 5-7 years.
- The other important setback for the Indian petrochemical industry is the price of feedstock which is comparatively higher in India as compared to China.
- India also lags behind as far as technological development is concerned.

Top Petrochemical Companies in India:

Though the Indian petrochemical industry is highly dominated by only a few players, however, there are a number of petrochemical companies in India, doing their share of business. Some of the top companies can be listed as below:

- Reliance Industries Ltd.
- Haldia Petrochemicals Ltd.
- Indian Oil Corporation.
- Gas Authority of India Limited.
- National Organic Chemical Industry Ltd.
- Bongaigaon Refinery and Petrochemicals Ltd.
- Manali Petrochemical Limited.
- I G Petrochemicals Limited.
- The Andhra Petrochemicals Limited.
- Tamilnadu Petroproducts Limited.

Key Segments in the Indian Petrochemical Industries:

Saturated paraffins are becoming important basic petrochemicals with the progress of petrochemical technology. Methane, Ethane and Propane are the most important members in this field. Otherwise they are used to produce other basic petrochemicals like olefin and butadiene.

The reason for their use as industrial and domestic fuel is due to their higher hydrogen/carbon ratio compared with olefins, dienes, naphthanes and aromatic, because the hydrogen to carbon ratio is measure of calorific value of fuel.

Methane :

Methane is the first and the lightest member of the petroleum hydrocarbon. It also contains the highest hydrogen/carbon ratio, and is therefore considered to be the premium fuel in industry. Natural gas is the main source of methane.

Refinery gases from various operations serve as the secondary source of methane and other aliphatic basic petrochemicals. Generally, in steam cracking and other milder hydrocracking operations methane is produced.

Ethane :

Like methane, ethane is also one of the constituents of natural gas. But its main sources are the refinery gases. Ethane is less reactive than ethylene and is used for the production of ethylene. Ethane will be increasingly used as a future basic petrochemical for manufacture of ethyl alcohol, acetaldehyde, acetic acid, and halogen derivative of paraffins.

Propane and Butane :

Propane is the principal constituent of liquefied petroleum gas (LPG). LPG is essentially a mixture of 95% propane and 5% butane. The principal uses of LPG are as the domestic fuel and feedstock for steam cracking for olefin production. Propane is also present in appreciable quantities in various refinery gases. Propane is used for production of ethylene and chemicals (34%), as a fuels (57%) and for other uses (9%).

4.1.1 Naphtha Steam Cracking Process

The steam naphtha cracker plant has two sections; a hot section and a cold section. The hot section consist of the preheater, furnace, indirect and direct quench system, primary fractionating column, compressor unit, caustic scrubber and drier.

The cold section consist of cold box, demetanizer, deethanizer, acetylene hydrogenator, ethane-ethylene separator, depropanizer, methyl-acetylene and propadiene hydrogenator, propane–propylene separator, debutanizer, butadiene, extractor and benzene extractor.

The naphtha feed enters the hot section of the cracker unit of the furnace to about 400°C by hot off gases from the furnace. It is mixed with steam and enters the radiation zone of the furnace to be heated to 800°C or higher very rapidly (residence time 0.2 to 0.4 sec) and pyrolized, and after that it is chilled quickly to about 400°C by indirect quenching in boiler to generate high pressure steam by direct quench oil obtained from the bottom fraction of the primary fractionating column.

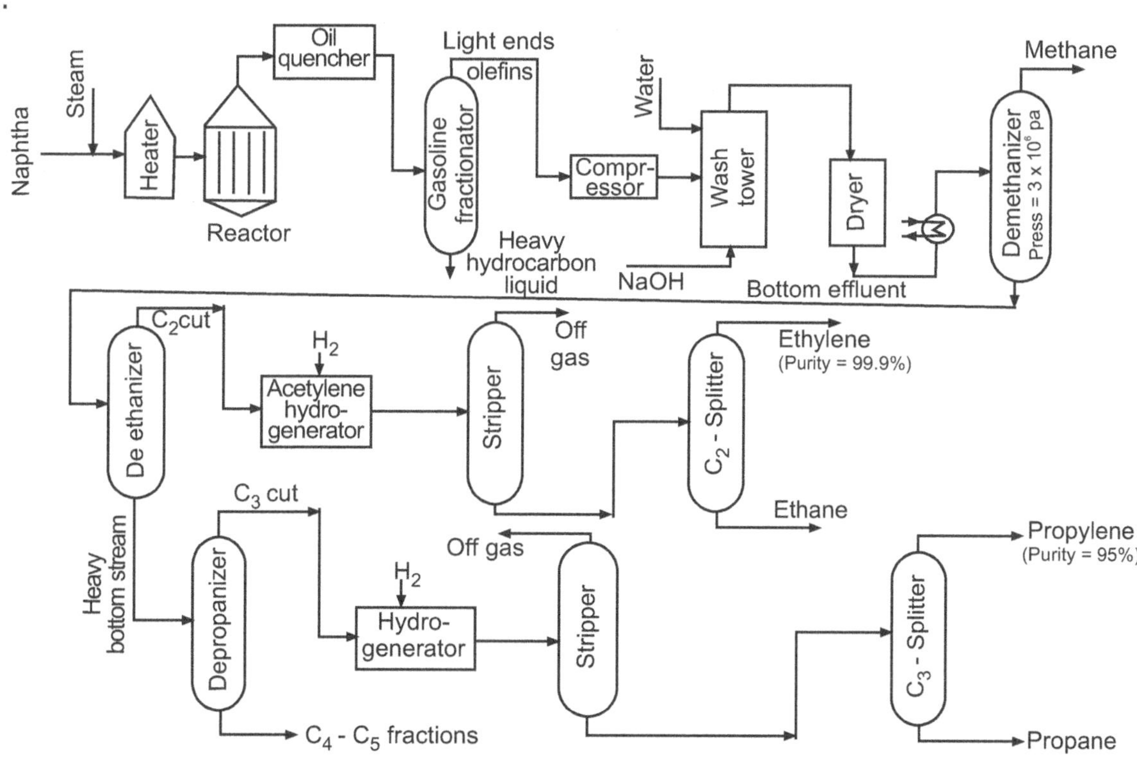

Fig. 4.1 : Steam naphtha cracking flow sheet

The cooled gas stream enters into a fractionating column where it is separated into light ends containing lighter olefins. The light ends undergo compression and condensation to be converted into liquid phase in several steps. These gases are desulfurized by scrubbing with ethanolamine and caustic soda and dried again compressed and cooled to a mixed gas liquid phase. This phase is sent to demethanizer (operating pressure is 3×10^6 pa) to separate methane as the top fraction and condensed to a liquid state at 100 - 140°C. The bottom effluent is sent to deethanizer where it is separated into the top stream of ethane and ethylene (C_2-cut) and heavy bottom stream. The top stream is sent to acetylene hydrogenator to convert acetylene present in this stream into ethylene over palladium catalyst at 40 - 80°C, and then to a column to separate ethylene as the light end product from ethane as bottom product, which may be recycled to cracker furnace.

The bottom fraction from deethanizer is again fractionated in the depropanizer into the top C_3 – fraction containing propane, propylene, methyl acetylene and propadiene, which is sent to hydrogenator to convert these unsaturated hydrocarbons into propylene. Hydrogenated stream is separated in a fractionating column into propylene (purity = 95%) as the top product and propane as the bottom product to be recycled to the furnace.

The bottom stream from the depropanizer is sent to the debutanizer to separate a C_4 cut containing 1,3-butadiene and other C_4 hydrocarbons as the top fraction. The C_4 cut is sent to the butadiene extraction unit to recover 1, 3 butadiene by extractive distillation.

The raw pyrolysis gasoline is further processed to recover motor gasoline, benzene, toluene, xylene (BTX) and other heavier aromatics.

4.2 C_1 HYDROCARBONS, PETROCHEMICALS FROM C_1 COMPOUND [W-13]

This group supplies about 25 weight % of all petrochemicals produced in the U.S. The processes studied in this section will cover the organic chemicals, particularly methanol, formaldehyde and halogenated hydrocarbons.

4.2.1 Production of Methanol [S-14]

(a) Properties of Methanol (CH3OH) :

Molecular Weight : 32.04

Melting Point : −97.8°C

Boiling Point : 64.7°C

(b) Methods of Production :

- Catalytic hydrogenation of carbon monoxide.
- Oxidation of LPG to give oxygenated hydrocarbon co-products of which methanol is a major fraction.
- Catalytic hydrogenation of carbon monoxide.

(c) Chemical Reaction :

(i) $CO + 2H_2 \longrightarrow CH_3OH$

(ii) Side reaction to methane

$$CO + 3H_2 \longrightarrow CH_4 + H_2O$$

$$2CO + 2H_2 \longrightarrow CH_4 + CO_2$$

(d) Temperature : 300 - 375°C

(e) Catalyst : Mixed catalyst of zinc, chromium, manganese, or aluminum oxides.

(f) Flow Sheet :

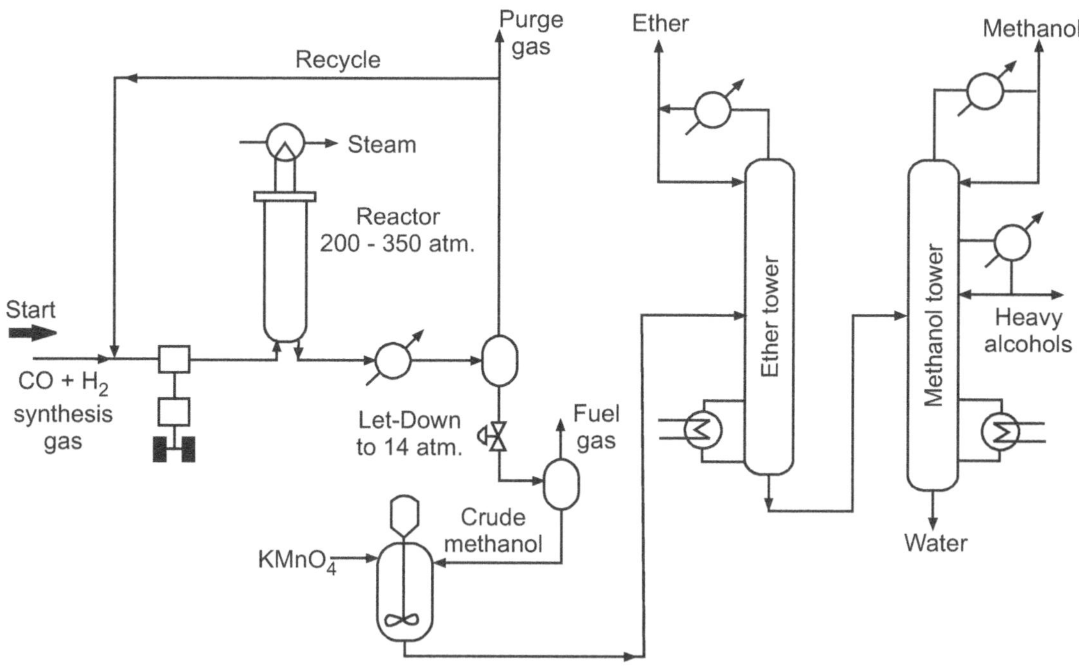

Fig. 4.2 : Methanol via synthesis gas route

(g) Process Description :

Hydrogen and carbon monoxide in a mole ratio of 2.25, is compressed to 3000 - 5000 psi, mixed with recycle gas, and fed to a high pressure converter. Internal preheat is usually employed. The reactor is copper lined steel and contains a mixed catalyst of zinc, chromium, manganese or aluminum oxides. The temperature is maintained at 300 - 375°C by proper space velocity and heat exchange design.

The exit gases are cooled by heat exchange with reactants, then with water. The methanol condenses under full operating pressure to maximize yields. The liquid methanol is depressurized, purified by permanganate to remove traces of ketones, aldehydes, and other such impurities, sent to a stripper to remove light ends such as dimethyl ether, and to a fractionator to separate methanol from the higher molecular weight compounds.

4.2.2 Production of Formaldehyde [W-13, 17; S-16, 17, 18]

(a) Properties of Formaldehyde (HCHO) :

Molecular Weight: 30.03

Melting Point : −118°C

Boiling Point : −19°C

(b) Methods of Production :

- Catalytic oxidation - dehydrogenation of methanol.
- Oxidation of methane or LPG to give oxygenated hydrocarbon co-products in which formaldehyde is separated with difficulty.
- Catalytic oxidation - dehydrogenation.

(c) Chemical Reaction :

(i) Oxidation

$$CH_3OH + \frac{1}{2} O_2 \longrightarrow HCHO + H_2O$$

(ii) Pyrolysis

$$CH_3OH \longrightarrow HCHO + H_2$$

(iii) Side reaction - complete combustion

$$CH_3OH + {}^3/_2\, O_2 \longrightarrow 2H_2O + CO_2$$

(d) Temperature : 450 - 600°C

(e) Catalyst : Silver or copper gauze or their oxides.

(f) Flow Sheet :

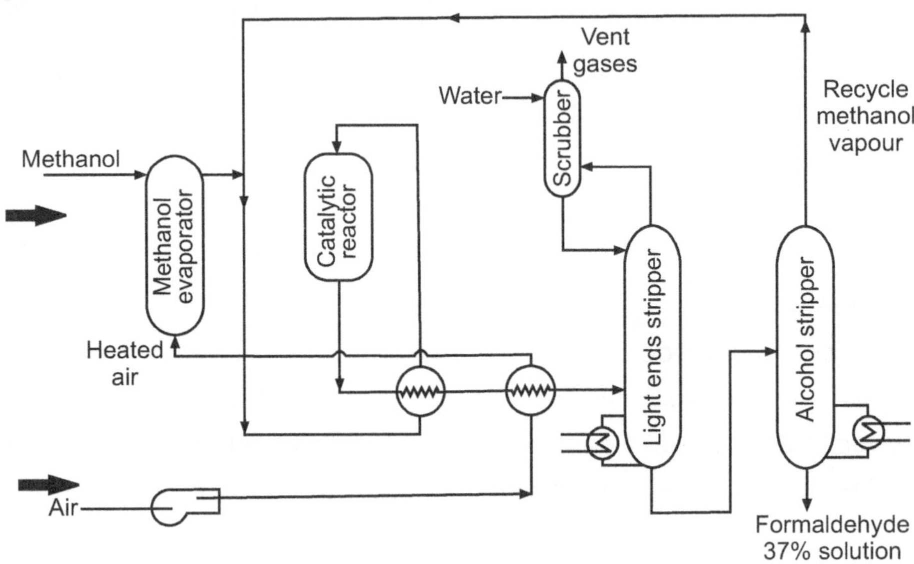

Fig. 4.3 : Formaldehyde from methanol

(g) Process Description :

Non-purified air, compressed to about 0.2 atm. gauge is preheated by heat exchange with reacting gases and then conveyed to a methanol evaporator. Conditions are set to keep the $CH_3OH:O_2$ ratio in the 30 - 50% range. The mixed gases are preheated, sent to a reactor where silver or copper gauze or their oxides are catalysts. Catalyst activity is controlled to maintain a balance between the endothermic dehydrogenation and exothermic oxidation reaction at reaction conditions of 450 - 600°C.

The product gases are absorbed in a water scrubber which is cooled by external circulation, and the fractionated to recover approximately 15% unreacted methanol for recycle. Ultimate yields of 85 - 90 wt.% based on methanol are typical.

4.2.3 Production of Chloromethanes [S-14]

The direct chlorination of methane leads to a series of chloromethanes from CH_3Cl to fully chlorinated CCl_4.

(a) Properties of Chloinated Methanes :

(i) Methyl Chloride (CH_3Cl)

Molecular Weight : 50.49

Melting Point : −97.7°C

Boiling Point : −24.1°C

(ii) Methylene Chloride (CH_2Cl_2)
Molecular Weight : 84.95
Melting Point : −97.6°C
Boiling Point : 40.1°C

(iii) Chloroform ($CHCl_3$)
Molecular Weight : 119.4
Melting Point : −63.5°C
Boiling Point : 61.2°C

(iv) Carbon Tetrachloride (CCl_4)
Molecular Weight : 153.84
Melting Point : −22.6°C
Boiling Point : 76.8°C

(b) Methods of Production :

- Direct thermal methane chlorination.
- CH_3Cl by reaction of methanol and HCl.
- CCl_4 by reaction of CS_2 and Cl_2.

(c) Chemical Reaction :

(i) $CH_4 + Cl_2 \longrightarrow CH_3Cl + HCl$

(ii) $CH_3Cl + Cl_2 \longrightarrow CH_2Cl_2 + HCl$

(iii) $CH_2Cl_2 + Cl_2 \longrightarrow CHCl_3 + HCl$

(iv) $CHCl_3 + Cl_2 \longrightarrow CCl_4 + HCl$

(d) Temperature : 370 - 410°C

(e) Catalyst : Silver or copper gauze or their oxides.

(f) Flow Sheet :

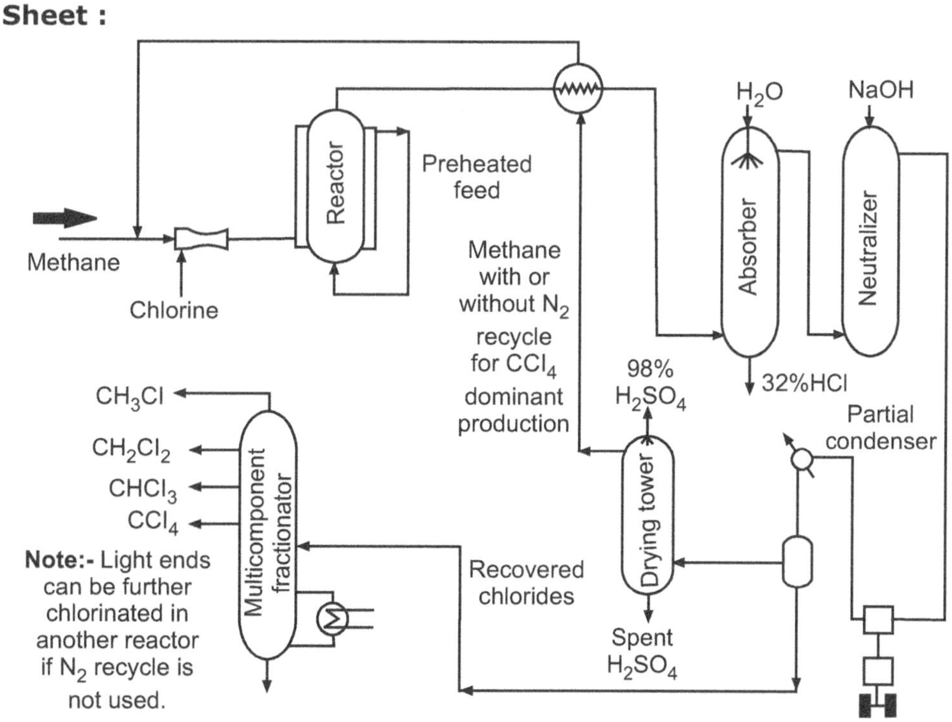

Fig. 4.4 : Direct chlorination of methane to produce methyl chloride, methylene chloride, chloroform and carbon tetrachloride

(g) Process Description :

Thermal chlorination of natural gas is accomplished by different cycles depending on the desired product ratio and the state of by-product HCl (dry gas or 30 - 32% muriatic acid).

Chlorination is carried out in an adiabatic furnace in which the exothermic heat evolved is recovered for preheating inlet gas to 280°C. The temperature is kept between 370°C and 410°C according to a selected cycle. The CH_4/Cl_2 ratio is fixed by the product ratio desired.

Hydrogen chloride is recovered in a water-spray absorption tower, giving 30 - 32% aqueous HCl directly. If dry HCl is desired, the azeotropic mixture HCl-H_2O is circulated through the tower, hence to a stripper, where dry HCl gas is the top effluent, the bottom being recirculated to the absorber.

Chlorinated gases from the absorber are scrubbed with caustic soda solution to remove HCl and CO_2 (the latter otherwise would accumulate as an inert gas in the recycle stream). Compression and drying with H_2SO_4 is followed by condensation and rectification to produce high purity products.

4.2.4 Production of Trichloroethylene

It represents one of the major chemicals produced from acetylene, of which a substantial fraction comes from natural gas. Other chemicals derived from acetylene are acrylonitrile.

(a) Properties of Trichloroethylene (C_2HCl_3) :

Molecular Weight : 131.40

Melting Point : −73°C

Boiling Point : 87.2°C

(b) Methods of Production :

- Vapour phase dehydrogenation of tetrachloroethane, using $BaCl_2$ catalyst.
- Liquid phase dehydrochlorination of tetrachloroethane, using milk of lime.

(c) Chemical Reaction :

(i) $C_2H_2 + 2Cl_2 \xrightarrow{FeCl_3} C_2H_2Cl_4$

(ii) $C_2H_2Cl_4 \xrightarrow[250\text{ - }300°C]{BaCl_2} C_2HCl_3 + HCl$

(d) Temperature : 250 - 300°C

(e) Catalyst : Ferric or antimony chloride (i.e. $FeCl_2$ or $BaCl_2$)

(f) Flow Sheet :

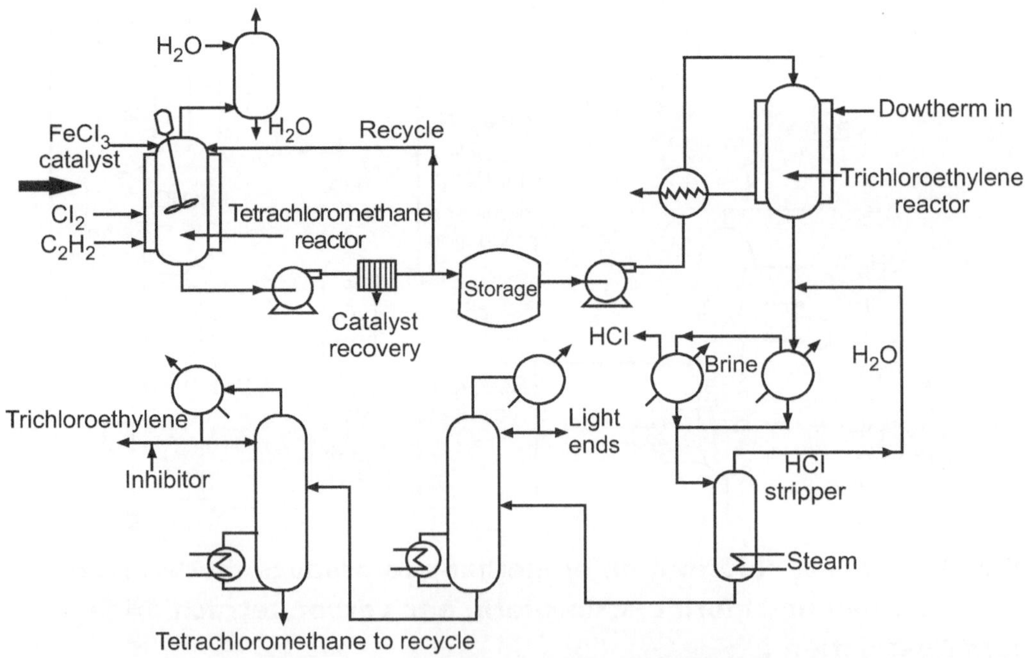

Fig. 4.5 : Trichloroethylene by vapour phase dehydrochlorination of tetrachloroetahne via acetylene-Cl_2

(g) Process Description :

Tetrachloroehane is produced by feeding chlorine and acetylene into a reactor containing recycle tetrachloroethane. Ferric or antimony chloride is the catalyst used in low concentration. The reaction is exothermic and is held at 80 - 100°C by external cooling. Product is removed and purified by distillation, or by settling filtration to remove catalyst.

The vapour phase dehydrochlorination is accomplished by preheat and reaction at 250 - 300°C over $BaCl_2$ catalyst deposited on porous carbon. The conversion to trichloroethylene is 90% per pass with little side reaction. Degassing of HCl with steam and a dual column distillation produces acid free grade "Trichlor".

4.3 ESTERIFICATION PROCESS [W-13; S-14]

4.3.1 Introduction

Esters are carboxylic acid derivatives with the general formula R-COOR′ (R, R′=H, alkyl or aryl). When a carboxylic acid is treated with a large excess of an alcohol, in presence of an acid catalyst, an ester is formed. The reaction is called acid catalysed esterification. The reaction between alcohols and carboxylic acids to make esters. It also looks briefly at making esters from the reactions between acyl chlorides (acid chlorides) and alcohols, and between acid anhydrides and alcohols.

The esterification process introduced by E. Fischer and A. Speier (1895) consists in refluxing acid and excess methanol or ethanol in presence of about 3% hydrogen chloride. The reaction is represented as:

$$RCOOH + \xrightleftharpoons{H^+ \; R'OH} RCOOR' + H_2O \qquad \dots (4.1)$$

Esterification is a chemical reaction that forms at least one ester (= a type of compound produced by reaction between acids and alcohols). Esters are produced when acids are heated with alcohols in a process called esterification. An ester can be made by an esterification reaction of a carboxylic acid and an alcohol.

Esterification is a widely employed reaction in the organic process industry.

Esters fall under a very wide category, ranging from aliphatic to aromatic with various substitutions and multifunctional groups, organic esters being valuable intermediates in the chemical industry. Generally, esters are derived from carboxylic acids. A carboxilic acid contains the –COOH group, and in an ester the hydrogen in this group is replaced by a hydrocarbon group of some kind. Esters are mostly used as plasticizers, solvents, perfumery and flavour chemicals and also as precursors to many pharmaceuticals, agrochemicals and fine chemicals.

The conventional esterification is an equilibrium reaction. For the stoichiometric mixture of acid and alcohol, equilibrium generally reaches 68% of conversion for the straight chain saturated alcohol. In order to obtain maximum yields, Le Chatlier's principle is followed, and the reaction is driven to the right hand side/forward direction, as follows:

Addition of one of the reactants in excess: The reaction usually reaches a point of equilibrium at 60% conversion, but in a small scale experiment a conversion of 60-80% can be achieved by use of a large excess of either acid or alcohol.

Removal of one of the products: Either ester or water formed is removed as soon as it is formed. Generally, suitable organic solvents are employed to remove the water formed during the reaction as a binary azeotrope or by employment of dehydrating agent such as anhydrous magnesium sulfate or molecular sieves.

4.3.2 Synthetic Routes to Esterification

To avoid equilibrium reaction, various precursors are used to achieve irreversible reaction, which involves the use of acyl halides or anhydrides in place of parent carboxylic acids. With acid halides though the reactions are rapid and irreversible, HCI produced as a byproduct is difficult to remove.

Further, an additional step is involved, where pyridine is used to remove HCI. Also such reactions are efficient for primary alcohols only. With secondary alcohols, the yields are lower and no reaction occurs with tertiary alcohols. Further, the aromatic acid chlorides are considerably less reactive when compared to aliphatic acid chlorides.

$$RCOCl + R'OH \longrightarrow RCOOR' + HCl \quad \text{... (4.2)}$$

When acid anhydrides are used, the reaction is slow and produces one mole of carboxylic acid in place of HCI. This process is widely accepted in industry for the esterification of polyhydroxy compounds such as carbohydrates, cellulose etc.

$$(RCO)_2O + R'OH \longrightarrow RCOOR' + RCOOH \quad \text{... (4.3)}$$

Other methods include the use of alkyl tert-butyl ether or by alkylation of salts of carboxylic acids with alkyl halides, transesterification, carbonylation of olefins, alcoholysis of nitriles etc.

The conventional catalyst used in esterification reactions is sulphuric acid, cited as a potential environmentally hazardous chemical, posing several disadvantages in industrial processes, which include wasting large amount of catalyst, corroding reactors, causing an acidic waste water, difficulty of catalyst recovery, and often lead to colouration of the product, in view of the deficiencies encountered, there is a global effort to replace the conventional homogeneous liquid acids by heterogeneous solid acids. Some of the solid acids used in esterification reactions are zeolites, niobic acid, solid super acids, heteropolyacids, pillared clays etc. They possess high catalytic activity and selectivity, do not corrode reaction vessels or reactors and finally, repeated use of solid acid catalyst from reaction mixture is easy and the product is free from colour and acid contamination.

4.3.3 Process

In this process, the alcohol and an acid are heated together in presence of sulfuric acid, and the reaction is driven to completion by removing the products as formed and employing an excess of one of the reagents. In the case of ethyl acetate, esterifications take place in a column which takes a ternary azeotrope overhead. Alcohol can be added to the condensed overhead liquid to wash out the alcohol, which is then rectified and returned to the column to react.

Amyl, butyl and isopropyl acetates are all made from acetic acid and the appropriate alcohols. All are useful lacquer solvents and their slow rate of evaporation prevents the surface of the drying lacquer from falling below the dew point, causing condensation on the film and a mottled surface appearance. This phenomenon is known as "blushing". Other esters are used in perfumery and in plasticizers. These include methyl salicylate, methyl anthranilate, diethyl-, dibutyl-, and di 2-ethylhexylphthalates.

Unsaturated vinyl esters for use in polymerization reaction are made by the esterification of olefins. The most important ones are vinyl acetate, vinyl chloride, acrylonitrile and vinyl fluoride. The addition reaction may be carried out in either the liquid, vapour or mixed phases. The choice depends upon the properties of the acid.

Vinyl acetate is manufactured by reacting ethylene with acetic acid in the vapour phase over a supported palladium catalyst.

$$H_2C-CH_2 + CH_3COOH + {}^1/_2\, O_2 \longrightarrow CH_3COOH = CH_2 + H_2O$$

The reaction is conducted in a fixed bed tubular reactor and is highly exothermic. With proper conditions, the only significant byproduct is CO_2 and enough heat is recovered as steam to perform the recovery distillation. Reaction is at 175 to 200°C under a pressure of 475 to 1000 kpa. To prevent polymerization, an inhibitor such as diphenylamine or hydroquinone is added.

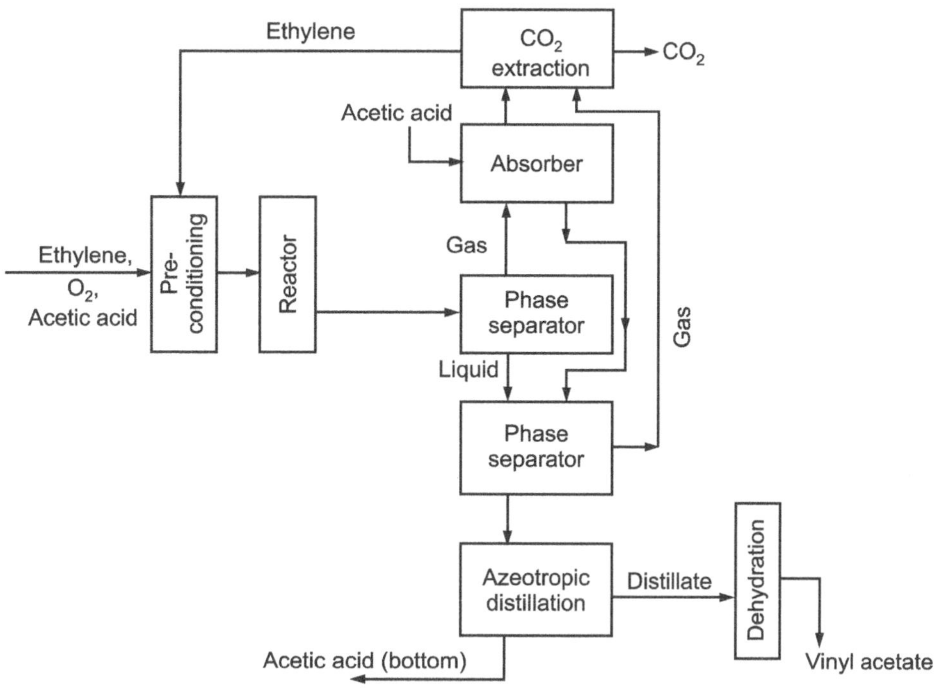

Fig. 4.6: Process of esterification

Esters of alkyl alcohol, e.g. dialkyl phthalate, are used as bifunctional polymerization monomers. They can be prepared by simple esterification of phthalic-anhydride with alkyl alcohol. Several acrylic esters, such as ethyl or methyl acrylates are also widely used and can be made from acrylic acid and the appropriate alcohol. The esters are more volatile than the corresponding acids.

4.4 C_2 HYDROCARBONS, PETROCHEMICALS FROM C_2 COMPOUND [W-13; S-16]

The C_2 chemicals are the most important of the petrochemicals as reflected by the 1971 projected requirements for India of 165,000 tons of ethylene plus 59,000 tons for acetylene as precursors.

4.4.1 Production of Ethanol [S-16, 17]

(a) Properties of Ethanol (C_2H_5OH) :

Molecular Weight: 46.07

Melting Point : −112°C

Boiling Point : 78.3°C

(b) Methods of Production :

Fermentation :

- From sucrose substrate
- From waste sulfite substrate of paper mills
- From starch substrate

Petroleum processing :

- Catalytic hydration of ethylene
- Esterification and hydrolysis of ethylene
- Oxidation of petroleum

(c) Chemical Reaction :

(i) Main reaction

$$C_{12}H_{22}O_{11} + H_2O \xrightarrow{\text{invertase}} 2C_6H_{12}O_6$$

$$C_6H_{12}O_6 \xrightarrow{\text{zymase}} 2C_2H_5OH + 2CO_2$$

(ii) Side reaction

$$2C_6H_{12}O_6 + H_2O \longrightarrow ROH + R'CHO$$

(d) Temperature : 20 - 35°C

(e) Catalyst : Yeast growth invertase and zymase catalytic enzymes.

(f) Flow Sheet :

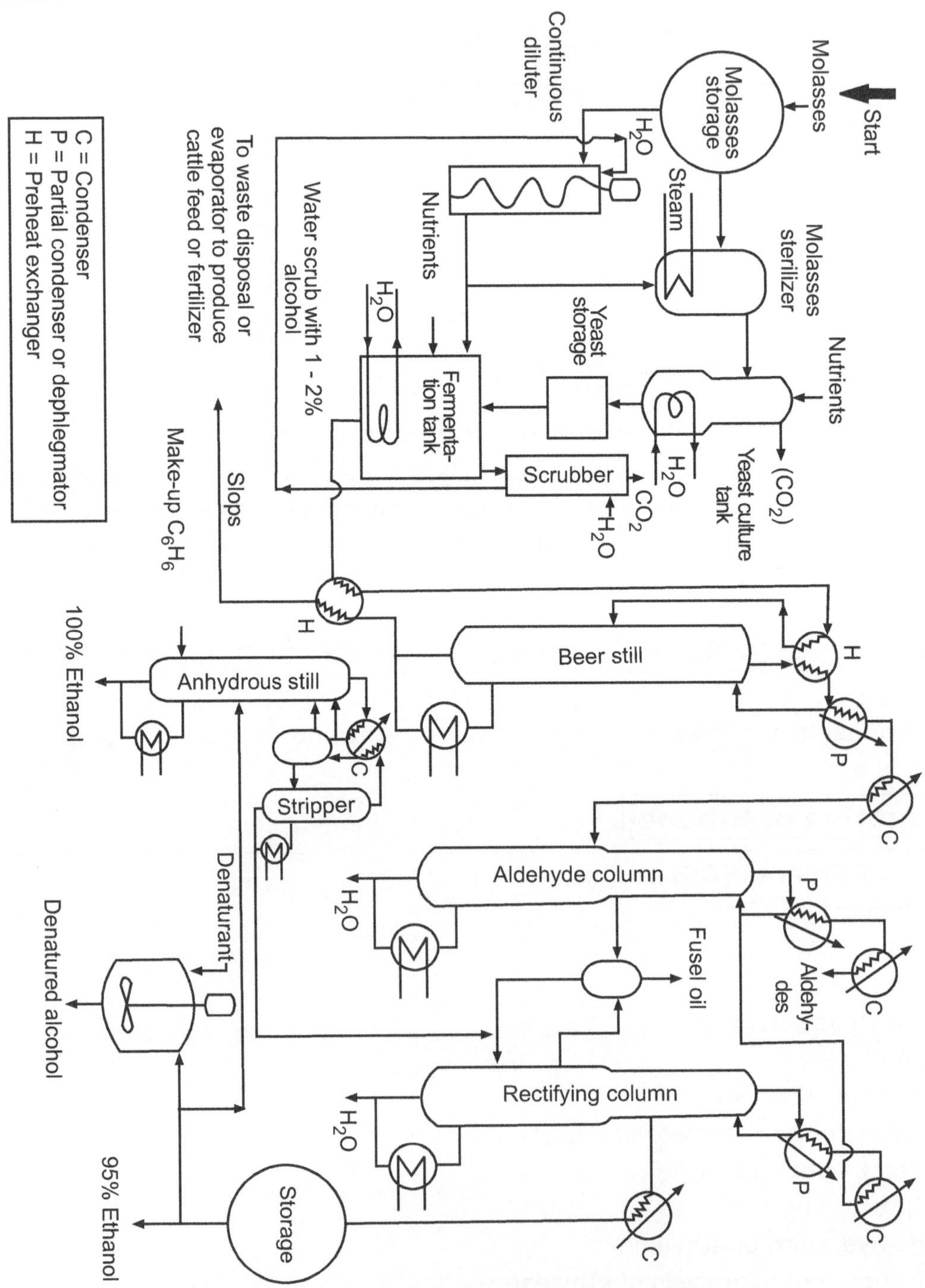

Fig. 4.7 : Ethyl alcohol by fermentation

(g) Process Description :

Molasses is diluted to a 10 - 15% sugar concentration and adjusted to a pH of 4 - 5 to support yeast growth which furnishes invertase and zymase catalytic enzymes. Nutrients such as ammonium and magnesium sulfate or phosphate are added when lacking in the molasses. This dilute mixture, called mash, is run into large wooden or steel fermentation tanks.

Yeast solution, grown by inoculating sterile mash, is added and fermentation ensues with evolution of heat which is removed via cooling coils. The temperature is kept at 20 - 30°C over a 30 - 70 hour period, rising near the end to 35°C. Carbon dioxide may be utilized as a by-product by water scrubbing and compressing; otherwise it is vented after water scrubbing.

Separation of the 8 - 10% alcohol in the fermented liquor called beer is accomplished by a series of distillations. In the beer still, alcohol (50 - 60% conc.) and undesirable volatiles such as aldehydes are taken off the top and fed to the aldehyde still. Alcohol is pulled off as a side-stream split to the rectifying column, the azeotropic alcohol-water mixture of 95% ethanol is taken off as a top side stream, condensed and run to storage tanks. When fusel oil recovery is practiced, side streams are drawn off near the bottom of the aldehyde and rectifying columns, and are separated by decantation. These higher molecular weight alcohols are sold directly for solvents or are fractionated to give predominantly amyl alcohol.

The bottom from the beer still, known as slops, are either discharged as waste or concentrated by evaporation to cattle feed depending on fuel and by-product sales economics.

4.4.2 Production of Ethylene Dichloride (1, 2-Dichloroethane) [W-17]

This compound is of interest as it is one of the intermediates for vinyl chloride monomer which polymerizes to polyvinyl chloride (PVC).

(a) Properties of Ethylene Dichloride ($ClCH_2 \cdot CH_2Cl$) :

Molecular Weight : 98.97

Melting Point : −35.3°C

Boiling Point : 83.7°C

(b) Methods of Production :

- Reaction of chlorine with ethylene in liquid or vapour phase.
- By-product of direct chlorination of ethane to ethyl chloride.
- By-product of chlorinated hydrocarbons.

(c) Chemical Reaction :

$$CH_2{=}CH_2 + Cl_2 \longrightarrow ClCH_2{\cdot}CH_2Cl$$

(d) Temperature : 45 - 50°C

(e) Catalyst : Traces of ferric chloride or ethylene dibromide.

(f) Flow Sheet :

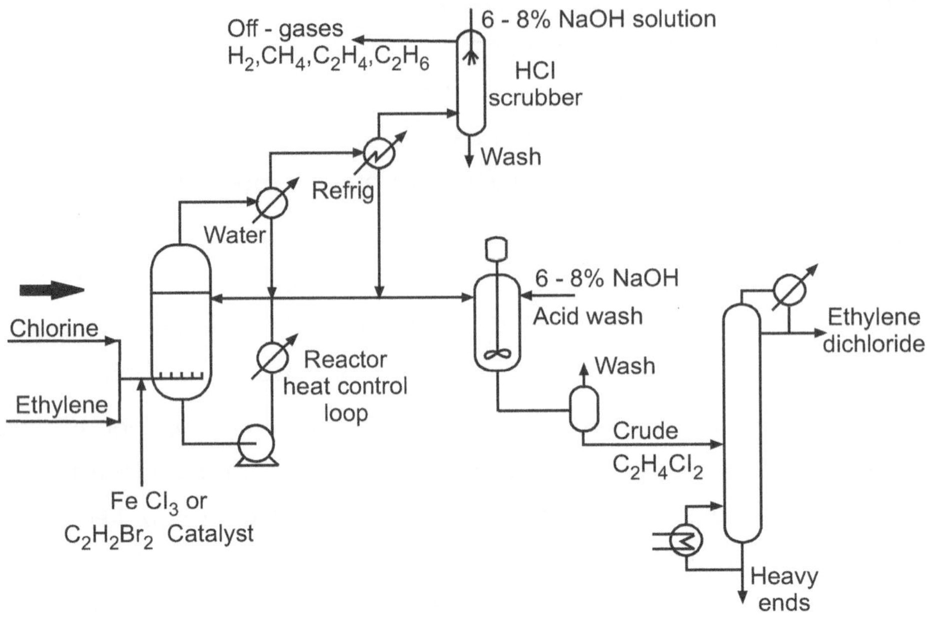

Fig. 4.8 : Ethylene dichloride production

(g) Process Description :

Ethylene (with or without C_2H_6, CH_4 and H_2 diluent) is mixed with chlorine and bubbled through a liquid phase reactor. Ethylene dichloride product serves as a reacting medium. Heat of reaction is controlled by external heat exchange and recycle, or by coil or jacket heat transfer, to hold the reactor at 45 - 50°C with a pressure of 1.5 - 2 atms. Traces of ferric chloride or ethylene dibromide as a catalyst gives 90 - 95% yield with little dimer formation. The gaseous products are cooled in two stages to strip the acid gas of ethylene dichloride. The liquid product is alkali washed and fractionated.

4.4.3 Production of Vinyl Chloride **[S-14]**

This compound is of interest as it is one of the intermediates for vinyl chloride monomer which polymerizes to polyvinyl chloride (PVC).

(a) Properties of Vinyl Chloride (CH_2=CHCl) :

Molecular Weight : 62.50

Melting Point : −153.8°C

Boiling Point : −13.81°C

(b) Methods of Production :

- Ethylene dichloride thermal pyrolysis.
- Acetylene-HCl reaction.
- Ethylene dichloride-caustic reaction.

(c) Chemical Reaction :

$$CH_2Cl.CH_2Cl \xrightarrow{500°C,\ 4\ atms.} CH_2{=}CHCl + HCl$$

(d) Temperature : 480 - 520°C

(e) Catalyst : Pumice or Charcol

(f) Flow Sheet :

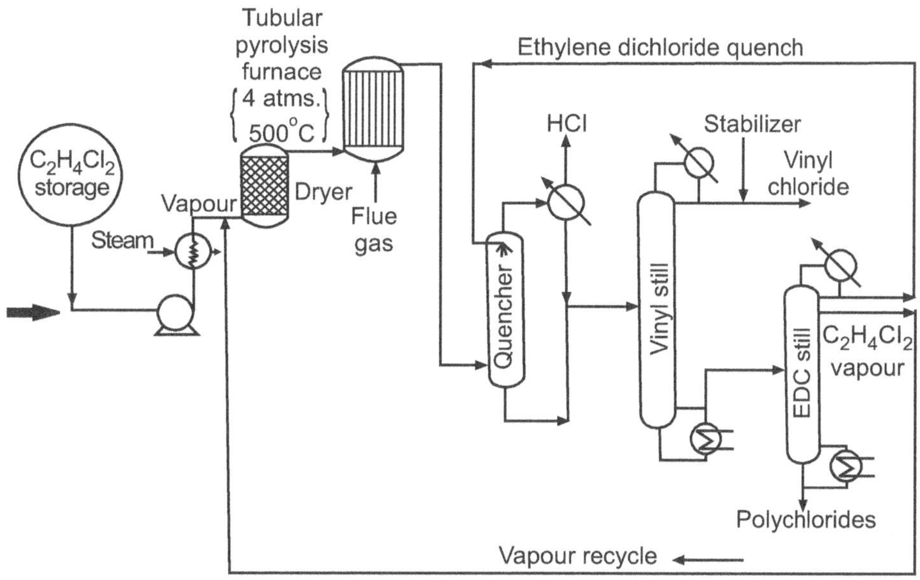

Fig. 4.9 : Vinyl chloride via ethylene dichloride pyrolysis

(g) Process Description :

Ethylene dichloride (EDC) vapours at 4 atms. are dried by silica gel and sent to a stainless steel tubular cracking furnace. This is externally flue gas fired and controlled at 480 - 520°C. The contact surface catalyst within the tubes is pumice or charcoal. The conversion per pass is around 50% and the ultimate yield is 90 - 96%. Spray quenching with cold EDC prevents back reaction. Uncondensed gases are sent to a surface heat exchanger to remove the balance of the EDC and vinyl chloride.

The non-condensibles containing HCl are either sent to the acetylene-HCl process in an adjacent process area or water scrubbed to recover HCl as muriatic acid.

4.4.4 Production of Ethylene Oxide **[W-13; S-16, 17, 18]**

(a) Properties of Ethylene Oxide(C_2H_4O) :

Molecular Weight : 44.05

Melting Point : −111.7°C

Boiling Point : 10.7°C

(b) Methods of Production :

- Direct oxidation of ethylene.
- Chlorohydrination of ethylene.

(c) Chemical Reaction :

$$CH_2{=}CH_2 + \tfrac{1}{2} O_2 \xrightarrow[\text{AgO}]{250 - 300^\circ C,\ 4 - 5\ \text{atms.}} C_2H_4O + \underset{\text{Side reaction}}{(CO_2 + H_2O)}$$

(d) Temperature : 250 - 300°C

(e) Catalyst : Silver oxide

(f) Flow Sheet :

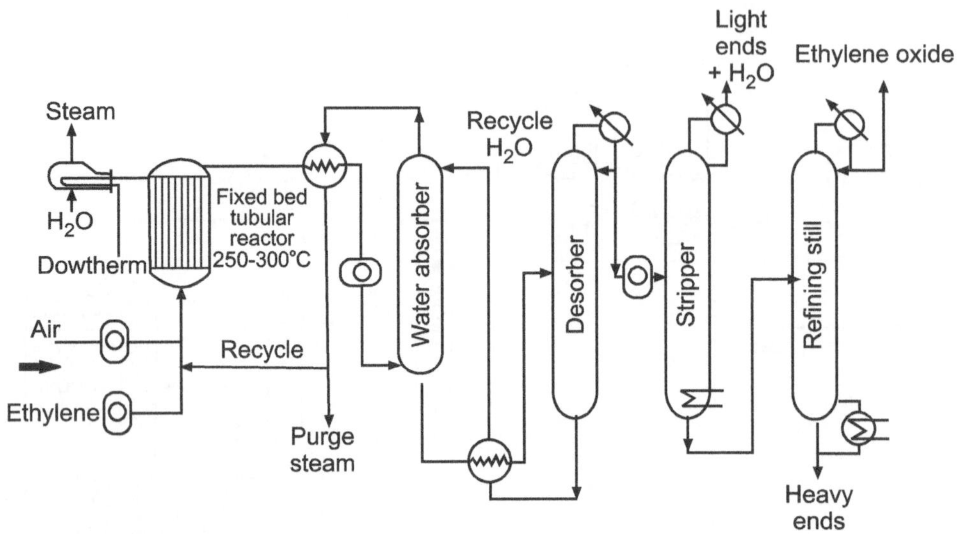

Fig. 4.10 : Ethylene oxide by air oxidation of ethylene

(g) Process Description :

Ethylene of 95 - 98% purity and air are compressed separately, mixed together giving 3 - 10% C_2H_4 volume concentration and passed over a catalyst of silver oxide on a porous inert carrier such as alumina. A side reaction suppressing agent such as ethylene dichloride is added to the feed to reduce the competitive oxidation reaction to CO_2 + H_2O.

The reaction is highly exothermic and is best carried out in a fixed bed tubular reactor in which heat transfer salt or dowtherm is pumped around the tube within the shell to maintain a 250 - 300°C temperature. Heat is recovered in a waste-heat steam boiler. A short residence time of 1 sec. in plug flow is the design with an ultimate yield of 60 - 70%.

The effluent gases from the reactor are water-washed under pressure. The absorbed ethylene oxide is sent to a packed bed desorber-fractionator tower and taken overhead. It still contains a large amount of water vapour plus some impurities. This stream is compressed to 4 - 5 atms. and fractionated twice to remove light ends, water, and high boiling polymers.

4.4.5 Production of Styrene [W-17; S-16, 17, 18]

(a) Properties of Styrene(C_8H_8) :

Molecular Weight : 104.1

Melting Point : −30.6°C

Boiling Point : 145.2°C

(b) Methods of Production :

- Dehydrogenation of ethylbenzene.
- Hydrogenation-dehydrogenation of acetophenone.

(c) Chemical Reaction :

(i) Alkylation of Benzene

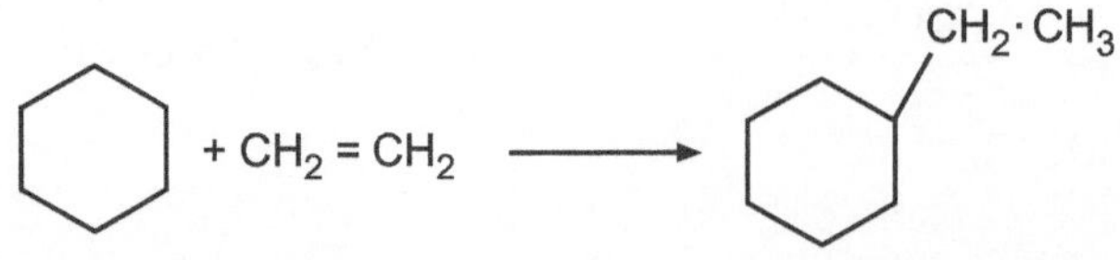

(ii) Dehydrogenation of ethylbenzene

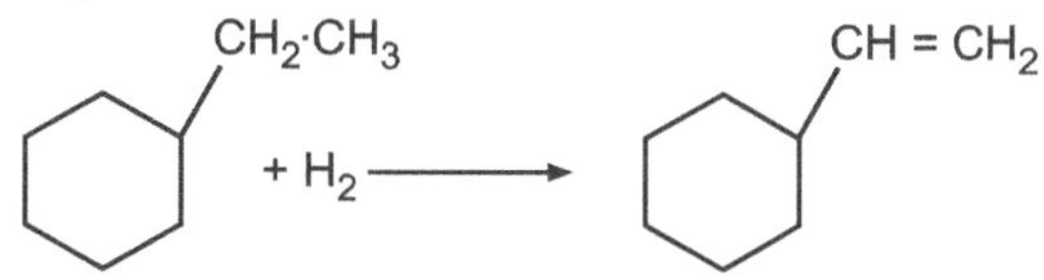

(d) Temperature : 200 - 500°C

(e) Catalyst : $AlCl_3$, promoted zinc, chromium, iron, or MgO on activated carbon, alumina, or bauxite.

(f) Flow Sheet :

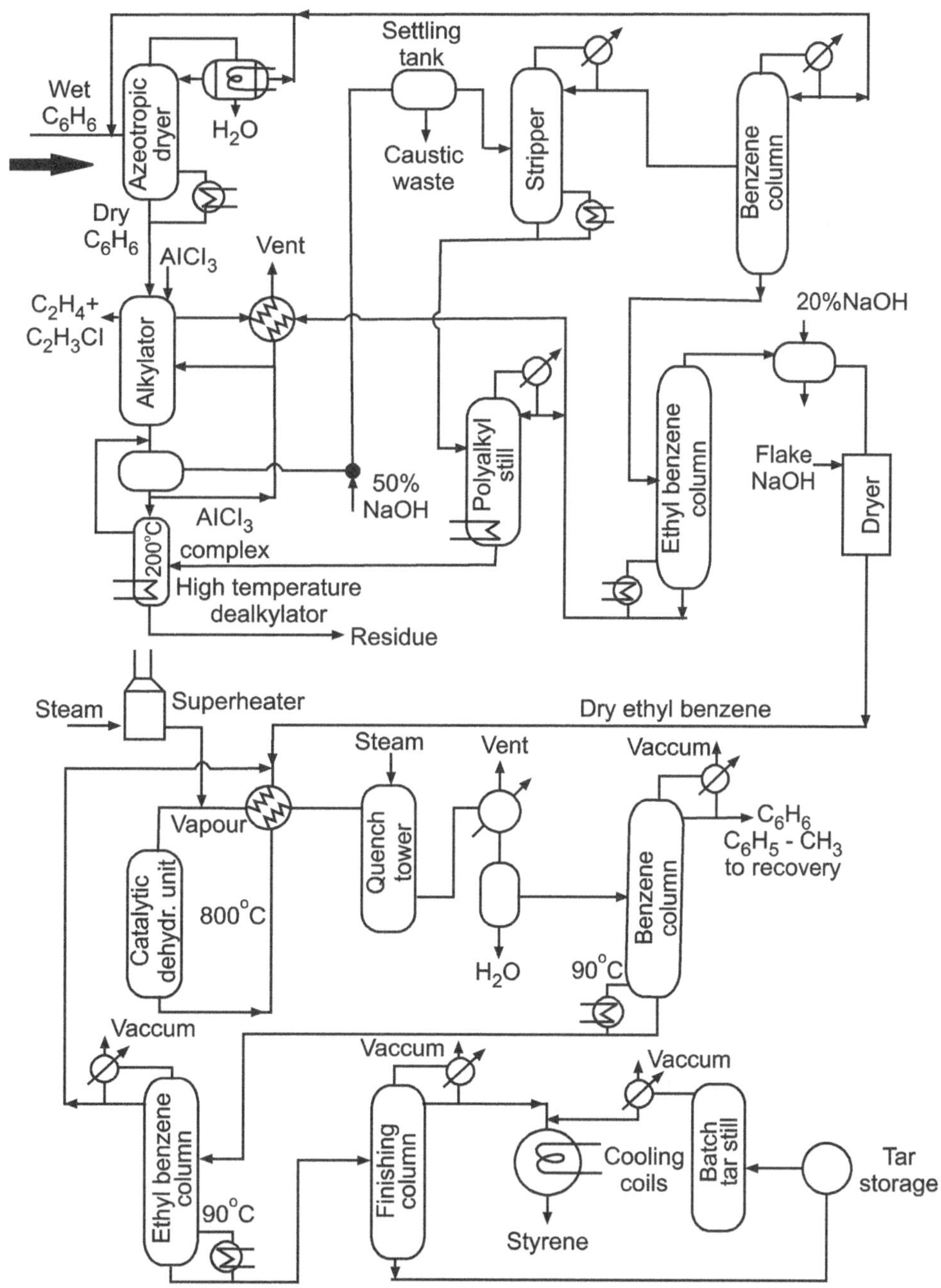

Fig. 4.11 : Styrene from benzene and ethylene

(g) Process Description :

The basic chemistry shows a two-step operation if the raw materials are more abundantly available ethylene and benzene compounds. Benzene is alkylated with ethylene by Friedel-craft method using aluminum chloride or other acid type catalyst. The resulting ethylbenzene is catalytically dehydrogenated in steam or excess benzene atmosphere to yield styrene.

The alkylation operation must be done under very dry conditions with high purity feedstock to avoid activity loss of catalyst. All benzene feed must be dried by azeotropic distillation. Ethyl chloride, which serves ultimately as a source of hydrogen and chlorine free radicals for catalyst promotion, is added to ethylene which is fed continuously along with benzene to the alkylation tower operating at 95°C and 1 atm. pressure. Granular $AlCl_3$ catalyst is fed continuously at the top of the reactor. The reactant mole ratio is 0.6 C_2H_4/1.0 C_6H_6 and no ethylene is recycled.

The alkylation tower is water cooled to control the exothermic reaction. The alkylate products are pumped to a cooler at 40°C where the aluminium chloride complex is separated and a split stream is fed to alkylator. The bleed-off fraction is pumped to a high temperature (200°C) dealkylator to break down the polyethylbenzenes to benzene and ethylbenzene which are returned to the system. The tar residue plus aluminum chloride is water extracted to recover 80 - 85% of $AlCl_3$.

Crude acidic ethylbenzene from the coolers is neutralized with 50% NaOH solution, stripped to remove polyethylbenzenes and the overhead sent to the benzene column which separates wet benzene from ethylbenzene. A final distillation is followed by 20% caustic wash, then drying by percolation in a flake caustic bed, produces 99% ethylbenzenes for dehydrogenation in yields averaging 95%.

The polyalkylbenzene must be removed as bottom product from atmospheric stripper to avoid formation of polyfunctional, highly reactive products during the dehydrogenation step. The stripper bottom fraction is distilled at 50 mm pressure with the overhead combined with bottoms from the ethylbenzene column and returned to the alkylation unit for dealkylation of these higher molecular weight compounds. The highest molecular weight polybenzenes as bottom from the vacuum still can only be split in the high temperature dealkylator.

Dehydrogenation of ethylbenzene is the step which produces styrene. Steam is injected with the hydrocarbon in a mole ratio of 15 H_2O/1 $C_6H_5C_2H_5$.

The mixed feed passes through the preheater to achieve an input temperature of 500°C to the reactor which operates at 800°C. The dehydrogenation catalyst is promoted zinc, chromium, iron, or MgO on activated carbon, alumina or bauxite. Conversion is 35 - 40% with an average yield of 91%.

Reaction product is cooled in the feed preheater, then by steam quenching. A water cooled condenser and separator operation yields crude styrene-ethylbenzene with impurities of 1% toluene, 0.7% benzene and 0.3% tar. Sulfur stabilizer is added at this point and the hydrocarbon mixture is passed through a series of vacuum distillation columns to allow the separation of impurities at low enough temperature to avoid styrene polymerization. Benzene and toluene are removed at 160 mm and a reboiler temperature of 90°C. Temperatures higher than this, cause autopolymerization of benzene. The second column at 35 mm and 90°C reboiler temperature separate styrene from ethylbenzene. The styrene, stabilized at this point with p-tertiary catechol, is vacuum distilled again to remove traces of tar and sulfur. The final overhead (99.8% or better styrene) is refrigerated at 10°C or less and either stored in insulated tanks or used immediately for polymerization.

4.5 C_3 HYDROCARBONS, PETROCHEMICALS FROM C_3 COMPOUND [S-16, 17]

4.5.1 Production of Acetyldehyde

It is one of the most important tonnage petrochemicals, being the raw material for acetic acid, acetic anhydride, acetate esters etc.

(a) Properties of Acetaldehyde (CH_3CHO) :

Molecular Weight : 62.50

Melting Point : −153.8°C

Boiling Point : −13.81°C

(b) Methods of Production :

- Hydration of acetylene.
- Oxidation of ethylene.

(c) Chemical Reaction :

$$C_2H_4 + \frac{1}{2} O_2 \longrightarrow CH_3CHO$$

(d) Temperature : 50 - 100°C

(e) Catalyst : Copper salt

(f) Flow Sheet :

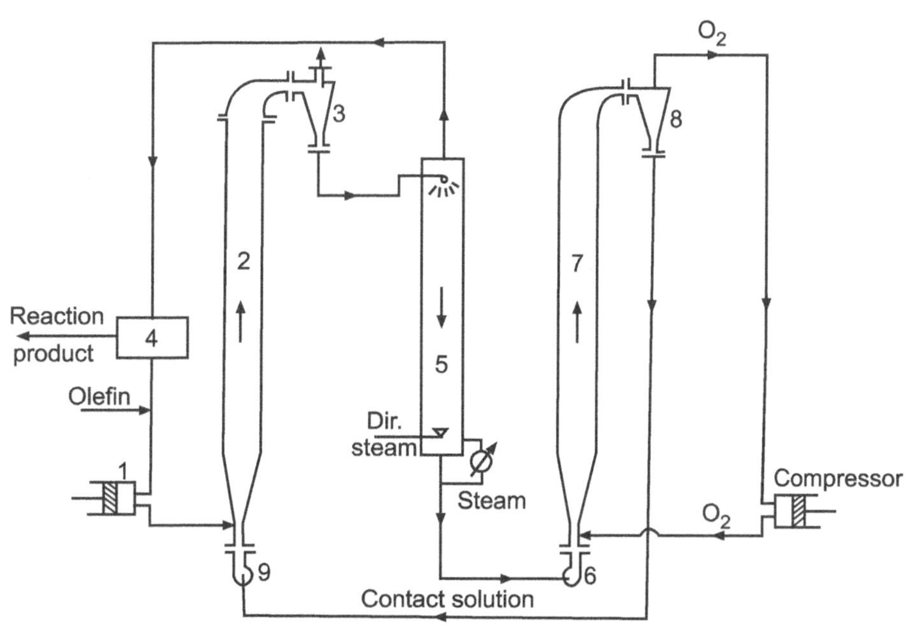

(1) - Compressor, (2) - Reactor, (3) and (8) - Cyclone separator, (4) Product separator, (5) Stripper, (6) and (9) - Pump, (7) - Regenerator

Fig. 4.12 : Acetaldehyde directly from ethylene

(g) Process Description :

In the early 1960's a new process was announced starting from lower cost ethylene as the raw material and producing acetaldehyde at lower cost according to the equation

$$C_2H_4 + \frac{1}{2} O_2 \longrightarrow CH_3CHO$$

Indeed the new process has been such an economic success that it immediately obsoleted the old process. No more acetylene-based plants have been built while overall acetaldehyde capacity has expanded greatly the world over.

The process operates in the presence of an aqueous liquid copper salt catalyst, promoted by a metal such as palladium. Thus, there are really two reactions, a hydration reaction

$$C_2H_4 + 2CuCl_2 + H_2O \longrightarrow CH_3CHO + 2CuCl + 2HCl$$

and a catalyst regeneration reaction

$$2CuCl + 2HCl + \frac{1}{2} O_2 \longrightarrow 2CuCl_2 + H_2O$$

which add up to give the net overall reaction

$$C_2H_4 + \frac{1}{2} O_2 \longrightarrow CH_3CHO$$

The process is operated at pressure below 50 atm. and at temperatures of 50 to 100°C. Typical reaction times range from 6 to 40 minutes. Tower type reactors gives a yields of acetaldehydes about 96%.

4.5.2 Production of Isopropanol [W-13]

(a) Properties of Isopropanol [$(CH_3)_2.CHOH$] :

Molecular Weight : 60.1

Melting Point : −89.5°C

Boiling Point : 82.5°C

(b) Methods of Production :

- Direct catalytic hydration of propylene.
- Hydration of propylene via sulfation and hydrolysis.

(c) Chemical Reaction :

(i) Sulfation

$$CH_3CH{=}CH_2 + H_2SO_4 \longrightarrow \underset{\text{Isopropyl acid sulfate}}{(CH_3)_2CH\,(OSO_3H)}$$

(ii) Hydrolysis

$$(CH_3)_2CH\,(OSO_3H) + H_2O \longrightarrow \underset{\text{Isopropanol}}{(CH_3)_2\cdot CHOH} + H_2SO_4$$

(iii) Ether formation – side reaction

$$[(CH_3)_2CH]_2\cdot(SO_3H) + H_2O \longrightarrow \underset{\text{Diisopropyl ether}}{[(CH_3)_2CHO]_2} + H_2SO_4$$

(d) Temperature : 20 - 30°C

(e) Flow Sheet :

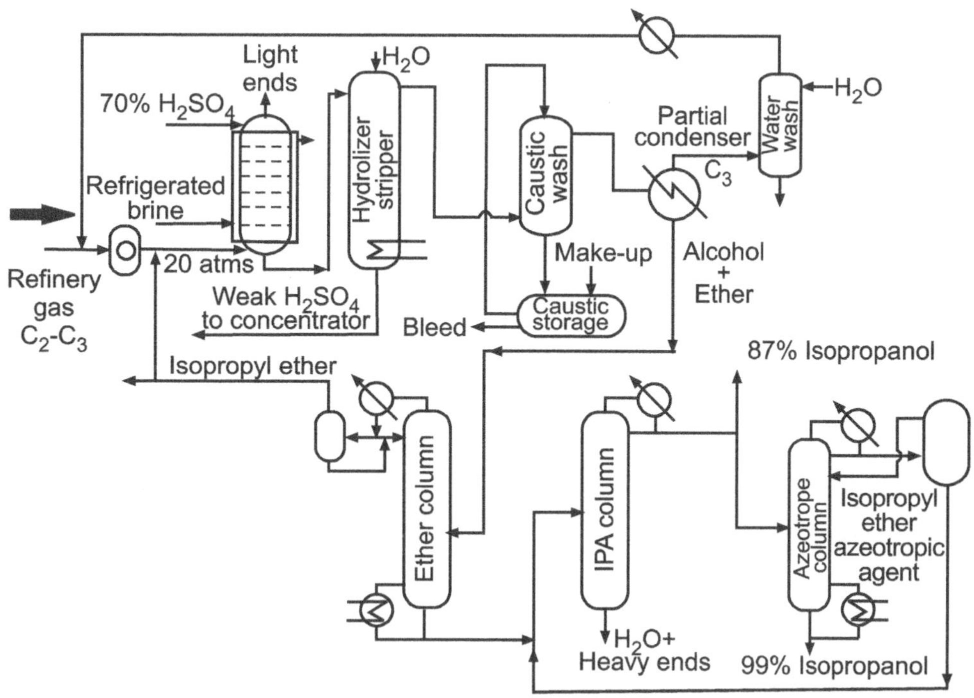

Fig. 4.13 : Isopropyl alcohol by indirect hydration of propylene

(f) Process Description :

Propylene, mixed with other C_2-C_3 hydrocarbon, is absorbed in 70 - 75% H_2SO_4 in a packed or sieve tray absorption tower in a series of agitated tower or in a series of agitated reactors operating at 20 - 25 atms. The strongly exothermic sulfation reaction is kept at 20 - 30°C by refrigerated brine circulation. The solution is hydrolyzed in a steam stripper which removes the organic phase as a vapour. The weak acid bottoms are sent to the H_2SO_4 concentration plant after neutralization in a caustic solution. The product vapour is fractionally condensed, with the non-condensible olefin fraction returned to the sulfation step.

The crude isopropanol is distilled to remove the more volatile isopropyl ether. A portion of the reflux is shunted to a decanter where ether rich top layer is pumped back to the sulfation step or dried for product sales. The water alcohol bottom layer is returned to the reflux column feed.

Ether column bottoms are pumped to the isopropanol column where an 87% IPA azeotrope is taken off the top. This can be sold as product or sent to a ternary azeotropic distillation column where isopropyl ether removes the water as overhead and the bottoms are 99% absolute isopropanol.

4.5.3 Production of Acetone

[W-13; S-14]

(a) Properties of Acetone (CH_3COCH_3) :

Molecular Weight: 58.08

Melting Point : −95.1°C

Boiling Point : 56.5°C

(b) Methods of Production :

- Catalytic dehydrogenation of isoproanol.
- Co-product of phenol process utilizing cumene.

- Co-product of glycerine-H_2O_2 process.
- Oxidation of butane.
- Oxidation of propylene.
- Fermentation of molasses.

(c) Chemical Reaction :

(i) $(CH_3)_2CHOH \longrightarrow CH_3CO\ CH_3 + H_2$

(ii) $(CH_3)_2CHOH + \frac{1}{2}\ O_2 \longrightarrow CH_3CO\ CH_3 + H_2O$

(d) Temperature : 500°C

(e) Catalyst : Copper or Brass

(f) Flow Sheet :

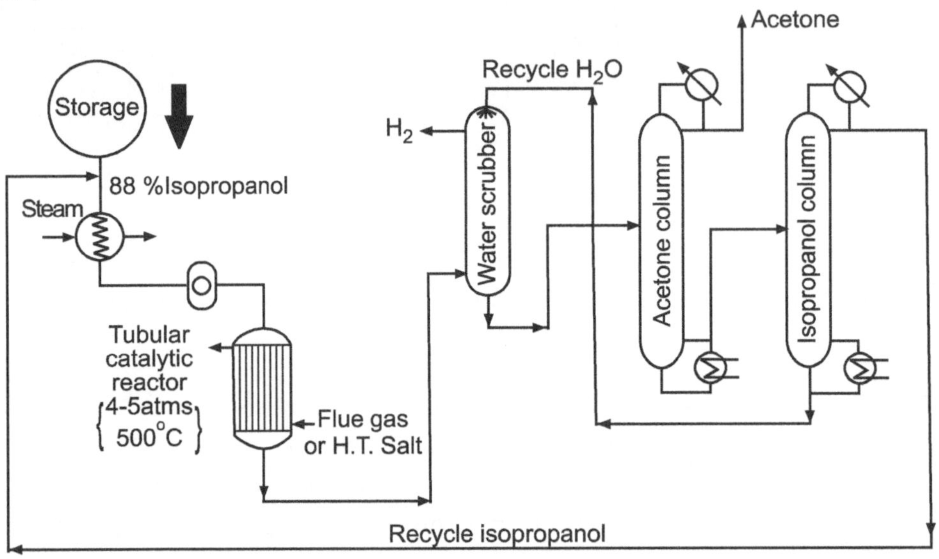

Fig. 4.14 : Acetone by dehydrogenation of isopropanol

(g) Process Description :

Isopropanol vapour compressed to 3 atms. is preheated by reactor effluent heat exchange and passed through a tubular catalytic reactor maintained at 500°C with the catalyst consisting of copper or brass deposited on porous carrier. The hot reaction gases passed through a water cooled condenser and hence into a water scrubber where final traces of isopropanol and acetone are removed from hydrogen. Condensate and water scrub liquor are fractionated to give product grade acetone overhead and dilute isopropanol bottoms. These are further fractionated to give 88% isopropanol recycle with the water recycle to the water scrubber.

4.5.4 Production of Cumene

[W-13; S-14]

(a) Properties of Cumene ($C_6H_5.C_3H_7$) :

Molecular Weight : 120.19

Melting Point : −96.9°C

Boiling Point : 152.5°C

(b) Methods of Production :

- Propylene alkylation of benzene.

(c) Chemical Reaction :

(i) Main reaction

$C_6H_6 + CH_3{\cdot}CH{=}CH_2 \longrightarrow C_6H_5{\cdot}C_3H_7$

Isopropyl benzene (cumene)

(ii) Side reaction

$$C_6H_6 + nCH_3{\cdot}CH{=}CH_2 \longrightarrow \underset{\text{Polyisopropyl benzene}}{C_6H_6\text{-}n{\cdot}(C_3H_7)n}$$

(d) Temperature : 250°C

(e) Catalyst : Impregnated H_3PO_4

(f) Flow Sheet :

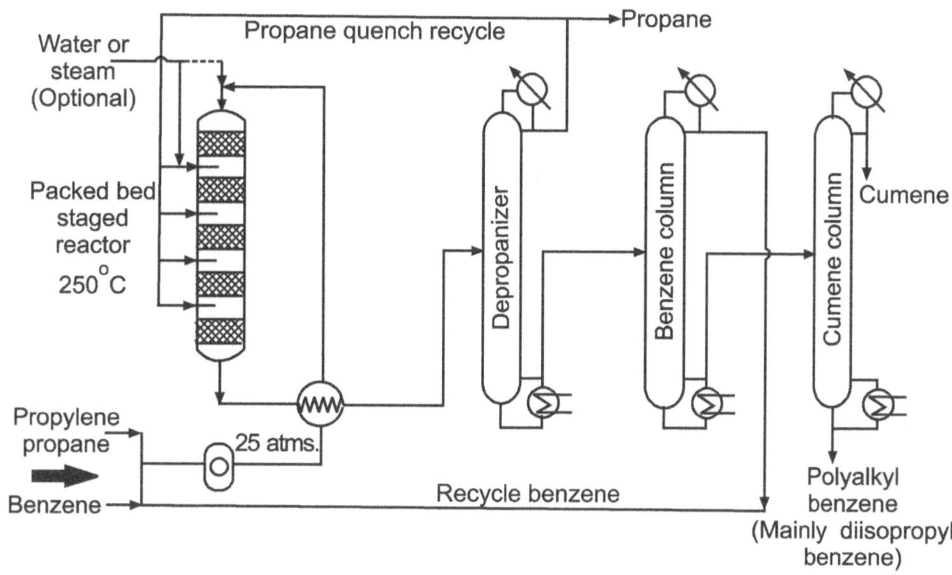

Fig. 4.15 : Cumene (Isopropyl benzene) via propylene alkylation of benzene

(g) Process Description :

Propylene-propane feedstock from refinery off-gases of a naphtha steam cracking plant is mixed with benzene and pumped at 25 atms. into the top of the reactor, packed stagewise with H_3PO_4 impregnated catalyst. The temperature is maintained at approximately 250°C by adding cold propane at each stage to absorb the heat of reaction.

The reactor effluent is depropanized and the propane split into quench or product streams. The depropanized bottoms are separated into benzene, cumene, and polycumenes in the remaining two stills.

4.5.5 Production of Propylene Oxide **[W-17; S-16, 17, 18]**

(a) Properties of Propylene Oxide ($CH_3-CH-CH_2$) :

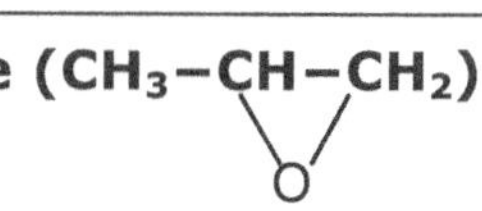

Molecular Weight : 58.08

Melting Point : −104.4°C

Boiling Point : 33.9°C

(b) Methods of Production :

- Chlorhydrin route.
- Direct oxidation.

(c) Chemical Reaction :

(i) Chlorhydrination

$$CH_3CH{=}CH_2 + HOCl \longrightarrow CH_3{\cdot}CHCl{\cdot}CH_2OH$$

(ii) Dehydrochlorination

$$2CH_3{\cdot}CHCl{\cdot}CH_2OH + Ca(OH)_2 \longrightarrow 2CH_3\text{-}\underset{\diagdown O \diagup}{CH\text{-}CH_2} + CaCl_2 + 2H_2O$$

(d) Temperature : 50°C

(e) Catalyst : Impregnated H_3PO_4

(f) Flow Sheet :

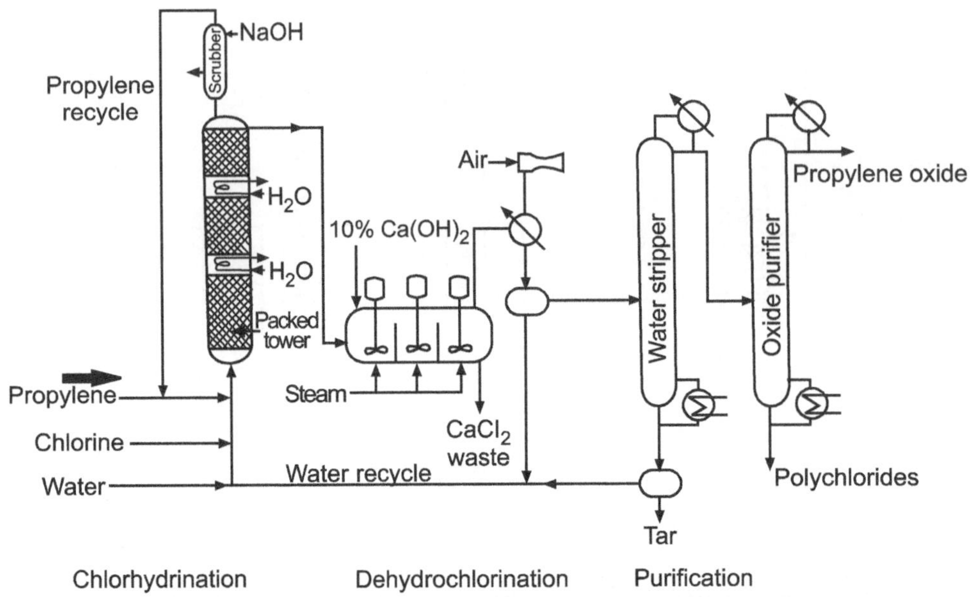

Fig. 4.16 : Propylene oxide via chlorhydrin

(g) Process Description :

Propylene, chlorine and water are introduced into the bottom of a packed tower where reaction forms the chlorhydrin. The reaction mechanism is formation of hypochlorous acid (HOCl or HClO) which reacts rapidly with propylene. The chlorine/propylene ratio of gaseous feed to the tower is chosen so that the liquid effluent contains 4 - 5% chlorhydrin. The reaction is exothermic and maximum tower temperature is held at 50°C. A conversion of 25 - 40% is attained, the ultimate yield of chlorhydrin is 87 - 90% with most of the by-product as propylene dichloride. Unreacted propylene is scrubbed with caustic to remove HCl, and then recycled.

Important Points

- Methane, Ethane, Propane and Butane are obtained from naphtha steam cracking process.
- **C_1 hydrocarbon compounds** : Methanol (CH_3OH), Formaldehyde (HCHO), Chloromethane [Methyl Chloride (CH_3Cl), Methylene Chloride (CH_2Cl_2), Chloroform ($CHCl_3$), Carbon Tetrachloride (CCl_4)], Trichloroethylene (C_2HCl_3).
- **Esterification process :** Esters are carboxylic acid derivatives with the general formula R-COOR′ (R, R′=H, alkyl or aryl). When a carboxylic acid is treated with a large excess of an alcohol, in presence of an acid catalyst, an ester is formed. The reaction is called acid catalysed esterification.
- **C_2 hydrocarbon compounds :** Ethanol (C_2H_5OH), Ethylene dichloride ($ClCH_2{\cdot}CH_2Cl$), Vinyl Chloride ($CH_2{=}CHCl$), Ethylene Oxide (C_2H_4O), Styrene (C_8H_8).
- **C_3 hydrocarbon compounds :** Acetaldehyde (CH_3CHO), Isopropanol [$(CH_3)_2{\cdot}CHOH$], Acetone (CH_3COCH_3), Cumene ($C_6H_5{\cdot}C_3H_7$), Propylene Oxide ($CH_3\text{-}CH\text{-}CH_2{\cdot}O$).

Practice Questions

1. Draw the flow sheet for manufacture of ethylene oxide.
2. Describe with neat flow sheet, the manufacturing process of phenol from Isopropanol.
3. Explain esterification process in detail.

MSBTE Questions and Answers (As Per G-Scheme)

Summer 2016

1. Draw flow diagram of ethylene oxide manufacture. **(4 M)**

Ans. Refer Section 4.4.4.

2. List out two chemical derived from C_1 and C_2 hydrocarbons. **(4 M)**

Ans. Refer Section 4.2 and 4.4.

3. Draw the flowsheet for manufacture of formaldehyde. **(4 M)**

Ans. Refer Section 4.2.2.

4. List out two chemicals derived from C_3 hydrocarbons with their uses. **(2 M)**

Ans. Refer Section 4.5.

5. Explain esterification process in refineries. **(4 M)**

Ans. Refer Section 4.3.

6. Explain the manufacturing of styrene with neat flow sheet. **(4 M)**

Ans. Refer Section 4.4.5.

7. Draw the flow sheet for manufacture of ethanol. **(4 M)**

Ans. Refer Section 4.4.1.

8. Draw the flow sheet for the manufacture of propylene oxide. **(4 M)**

Ans. Refer Section 4.5.5.

Summer 2017

1. Write the chemical reaction of the following with standard condition. **(4 M)**
   (i) Manufacture of methanol.
   (ii) Manufacture of formaldehyde.

Ans. Refer Sections 4.2.2 and 4.2.2.

2. Describe manufacture of ethylene oxide with well labelled flow diagram. **(6 M)**

Ans. Refer Section 4.4.4.

3. Describe manufacture of methanol with well labelled flow diagram. **(4 M)**

Ans. Refer Section 4.2.1.

4. Name any four petrochemicals derived from C_1 hydrocarbon with their one application. **(4 M)**

Ans. Refer Section 4.2.

5. Describe manufacture of formaldehyde with well labelled flow diagram. **(4 M)**

Ans. Refer Section 4.2.2.

6. Name any four petrochemicals derived from C_2 hydrocarbon with their one application. **(4 M)**

Ans. Refer Section 4.2.

7. Give two uses each of following petrochemicals. **(4 M)**
   (i) Formaldehyde
   (ii) Ethylene oxide
   (iii) Acetaldehyde

Ans. Refer Section 4.2.2, 4.4.4, 4.5.1.

8. Name any two physical properties of the following petrochemicals. **(4 M)**
   (i) Methanol

Ans. Refer Section 4.2.1 (a).
   (ii) Formaldehyde

Ans. Refer Section 4.2.2 (a).
   (iii) Ethanol

Ans. Refer Section 4.4.1 (a).
   (iv) Propylene oxide

Ans. Refer Section 4.5.5 (a).

9. Explain uses of : (i) Styrene, (ii) Ethanol. **(4 M)**

Ans. Refer Sections 4.4.5 and 4.4.1 respectively.

Winter 2017

1. List out two chemicals derived from C_1 and C_2 hydrocarbons. State 2 uses of each product. **(4 M)**

Ans. Refer Section 4.2 and 4.4.

2. With a neat flowsheet explain the manufacture of formaldehyde. **(4 M)**

Ans. Refer Section 4.2.2.

3. Give the reaction involved in the formation of ethylene oxide and styrene. **(4 M)**

Ans. Refer Section 4.4.2 and 4.4.5.

4. Draw the flowsheet for the manufacture of methanol. **(4 M)**

Ans. Refer Section 4.2.1.

5. Describe the manufacturing process of propylene oxide with neat flowsheet. **(4 M)**

Ans. Refer Section 4.5.5.

6. List the name of four petrochemicals derived from C_3 hydrocarbons. Give any one use each. **(4 M)**

Ans. Refer Section 4.5.

Summer 2018

1. Draw flow diagram for the manufacture of methanol. **(4 M)**

Ans. Refer Section 4.2.1.

2. Write reactions involved in the manufacture of styrene from benzene and ethylene. **(4 M)**

Ans. Refer Section 4.4.5.

3. Explain the manufacturing of formaldehyde with neat diagram. **(4 M)**

Ans. Refer Section 4.2.2.

4. Explain the manufacture of ethylene oxide. **(4 M)**

Ans. Refer Section 4.4.4.

5. List the chemicals derived from C_1 and C_2 hydrocarbons with their uses. **(4 M)**

Ans. Refer Section 4.2 and 4.4.

6. Describe the manufacture of propylene oxide with neat flow diagram. **(8 M)**

Ans. Refer Section 4.5.5.

Chapter **5...**

PETROCHEMICALS FROM C_4 AND AROMATICS

Weightage of Marks = 10, Teaching Hours = 08

Syllabus

5.1 Chemicals from C_4 and Manufacturing of Butadiene and Methyl Tertiary Butyl Ether

5.2 List of chemicals from aromatics

5.3 Production and separation of Benzene Toluene and Xylene

5.4 Manufacture of aniline from phenol

5.5 Manufacture of benzoic acid from toluene

About this Chapter

At the end of this chapter, students will be able to:

- State the chemical obtained from specified fraction of petrochemical C_4 or aromatics.
- Describe with flow sheet the manufacturing of specified of petrochemical C_4 or aromatics.
- Explain with sketch manufacturing of specified petrochemical C_4 or aromatics.
- Explain Udex process for recovery of BTX from the specified reformats gasoline.

5.1 INTRODUCTION

Saturated paraffins are becoming important basic petrochemicals with the progress of petrochemical technology. The main source of butane is the overhead flue gas obtained during atmospheric distillation of crude oil. It is also present in refinery gas streams.

The largest use of n-butane is in gasoline. This is usually directly blended to increase volatility without reducing the octane rating of the blended gasoline. However environmental protection agency has proposed to reduce the volatility of gasoline, which will adversely affect the use of n-butane.

5.2 C_4 HYDROCARBONS, PETROCHEMICALS FROM C_4 COMPOUND

5.2.1 Production of Butadiene

[S-14, 16, 17, 18; W-17]

(a) Properties of Butadiene ($CH_2=CH{\cdot}CH=CH_2$) :

Molecular Weight : 54.09

Melting Point : −108.9°C

Boiling Point : −4.41°C

(b) Methods of Production :

- Dehydrogenation of butane.
- Dehydrogenation of butylene.
- Dehydrogenation-dehydration of ethanol.
- Steam cracking of hydrocarbons.

(c) Chemical Reaction :

(i) Main reaction

$$C_4H_{10} \longrightarrow CH_2 = CH \cdot CH = CH_2 + 2H_2$$

(ii) Side reaction

$$C_4H_{10} \longrightarrow C_4H_8 + H_2$$

(d) Temperature : 550 - 650°C

(e) Catalyst : Furfural

(f) Flow Sheet :

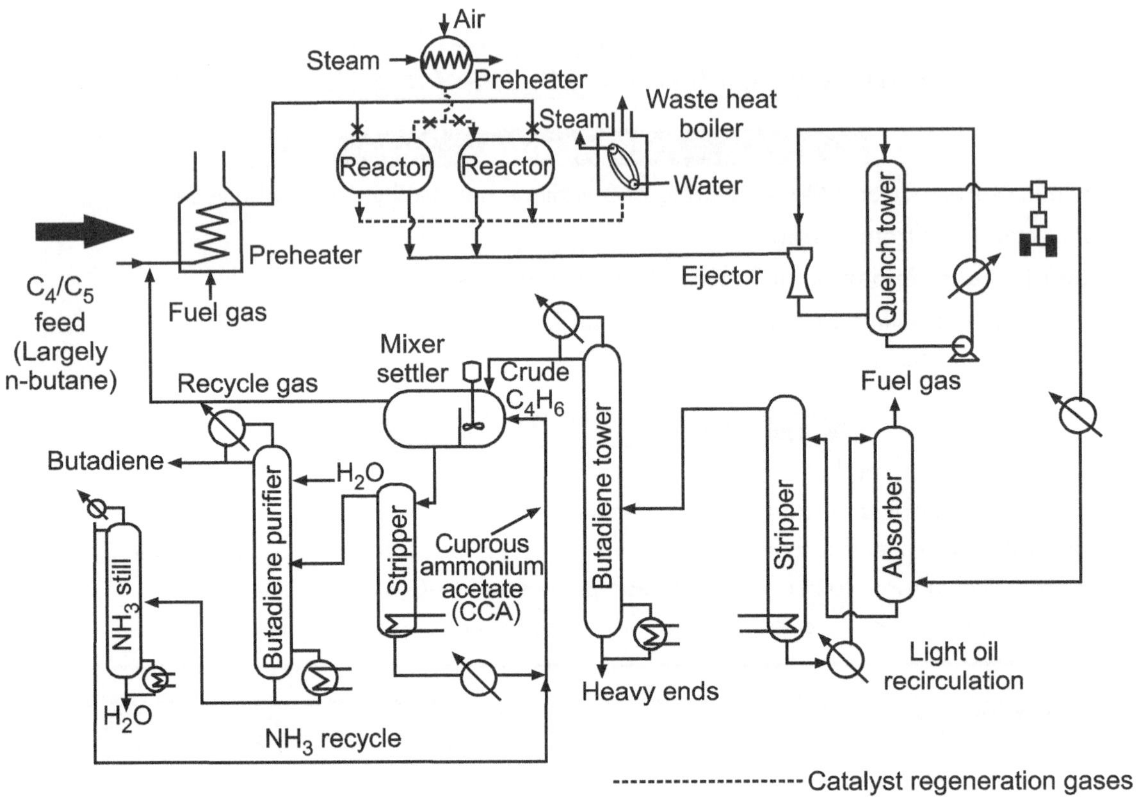

Fig. 5.1 : Butadiene via dehydrogenation of butane

(g) Process Description :

A refinery gas of C_4/C_5 containing predominantly n-butane with some isopentane is mixed with recycle gas and preheated to reaction temperature prior to contact with a catalyst in a fixed bed, regenerative-heating reactor system. A pair of reactors forms an adiabatic cycle with the heat of reaction required during the 5-15 minutes "make" period equal to that supplied by the combustion of carbon deposit on the catalyst during the regenerative period. The temperature of reaction at the start of the "make" period is 650°C, dropping to 550°C at the end before switching to regeneration. The pressure is low, 120 - 150 mm absolute, to force the reaction to the right.

The product gases are oil-quenched, compressed, cooled and separated from the light ends by absorption in naphtha followed by stripping. The overhead is fractionated to yield crude butadiene at the top which is purified by (1) absorption using cuprous ammonium acetate (CAA), (2) extractive distillation with furfural, or (3) azeotropic distillation with ammonia.

The more common absorption process involves contact of the close boiling butadiene-butene fraction with lean CAA solution which dissolves butadiene. A desorption step at higher temperature is followed by distillation, compression and liquefaction of butadiene to give 98.99% product purity. Ammonia is recovered in the distillation tower by water addition; it is then separated as substantially anhydrous NH_3 by fractionation.

5.2.2 Production of MTBE

[S-14, 16, 17, 18; W-13, 17]

Methyltertiary-butyl ether (MTBE) is a gasoline additive used to increase octane number that is produced from methanol and isobutylene. Methanol is purchased but the isobutylene is obtained from a refinery stream. The stream contains 23% isobutylene, 20% 1-butene and 57% 2-butenes and can be supplied at a maximum rate of 500 kmol/hr. Only isobutylene reacts with methanol.

MTBE is almost exclusively used as a fuel component in fuel for gasoline engines. It is one of the group of chemicals commonly known as *oxygenates* because they raise the oxygen content of gasoline. It is used as a reaction medium and extraction solvent to replace methylene chloride, aromatics and other ethers.

(a) Properties of MTBE ($C_5H_{12}O$) :

Molecular Weight : 88.15

Melting Point : −109°C

Boiling Point : 55.2°C

(b) Methods of Production :

- Catalytic etherification of isobutylene.

(c) Chemical Reaction :

$(CH_3)_2C{=}CH_2 + CH_3OH \longrightarrow (CH_3)_3C\text{-}O\text{-}CH_3$

(d) Temperature : 45 - 60°C

(e) Catalyst : Cation exchange resin.

(f) Flow Sheet :

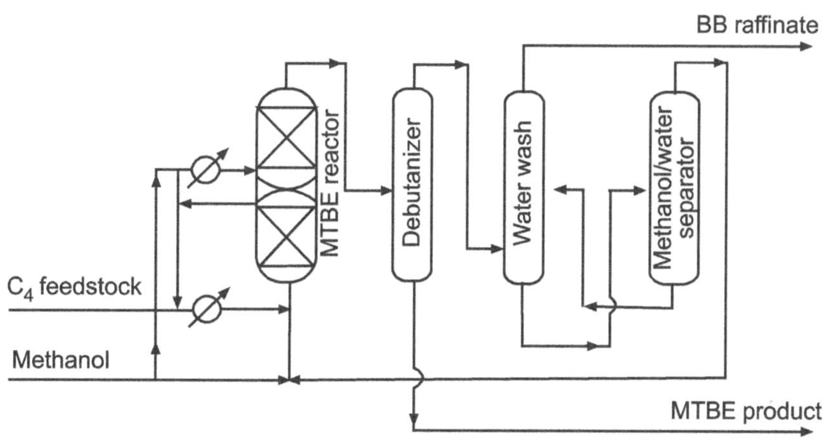

Fig. 5.2 : MTBE via catalytic etherification

(g) Process Description :

The technology features a two-stage reactor system of which the first reactor is operated in the recycle mode. With this method, a slight expansion of the catalyst bed is achieved which ensures very uniform concentration profiles within the reactor and, most important, avoids hot spot formation. Undesired side reactions, such as the formation of dimethyl ether (DME), are minimized.

The reactor inlet temperature ranges from 45°C at start-of-run to about 60°C at end-of-run conditions. One important factor of the two stage system is that, the catalyst may be replaced in each reactor separately, without shutting down the MTBE unit.

The catalyst used in this process is a cation-exchange resin and is available from several catalyst manufacturers. Isobutene conversions of 97% are typical for FCC feedstocks. Higher conversions are attainable when processing steam-cracker C_4 cuts, which contain isobutene concentrations of 25%.

MTBE is recovered as the bottom product of the distillation unit.

The methanol-rich C_4 distillate is sent to the methanol-recovery section.

Water is used to extract excess methanol and recycle it back to process.

The isobutene-depleted C_4 stream may be sent to a raffinate stripper or to a molsieve-based unit to remove other oxygenates such as DME, MTBE, methanol and tert-butanol.

Very high isobutene conversion, in excess of 99%, can be achieved through a debutanizer column with structured packings containing additional catalyst. This reactive distillation technique is particularly suited when the raffinate-stream from the MTBE unit will be used to produce a high-purity butene-1 product. For a C_4 cut containing 22% isobutene, the isobutene conversion may exceed 98% at a selectivity for MTBE of 99.5%.

5.2.3 Production of Butyl Acetate

(a) Properties of Butyl Acetate ($C_6H_{12}O_2$) :

Molecular Weight : 116.16

Melting Point : −74°C

Boiling Point : 126°C

(b) Methods of Production :

- By reacting butylene with acetic acid.

(c) Chemical Reaction :

$$C_4H_{10} + CH_3COOH + \frac{1}{2} O_2 \longrightarrow C_6H_{12}O_2 + 2H_2O$$

(d) Temperature : 175 - 200°C

(e) Catalyst : Palladium

(f) Process Description :

Butyl acetate is manufactured by reacting ethylene with acetic acid in the vapour phase over a supported palladium catalyst.

The reaction is conducted in a fixed bed tubular reactor and is highly exothermic. With proper conditions, the only significant by-product is CO_2. Enough heat is recovered to perform the recovery distillation. Reaction is at 175 to 200°C under a pressure of 1000 kPa. To prevent polymerization, an inhibitor such as diphenylamine is added. Units for this process are quite large, some producing 50×10^6 kg/year.

An older process reacts butylene with acetic acid in the liquid phase. Most butyl acetate is now produced from butylene.

5.3 AROMATIC FRACTIONS

5.3.1 Chemicals from Aromatics [S-17, 18; W-17]

The principal aromatics for chemical synthesis are benzene (B), toluene (T), xylene (X) and naphthalene.

Coal carbonization is used for metallurgical coke production. The volatiles from the coking operation are stripped of tar by electrostatic precipitators, scrubbed with dilute H_2SO_4 (10 - 20%) to remove ammonia. The remaining crude benzene fraction is absorbed in a high boiling stable oil such as creosote. This oil has approximate composition of 60% B, 15% T, 7% X, 4% N and the solvent naphtha 7% and is fractionated to separate these components.

The production of aromatic raw materials from petroleum is derived from (1) solvent extraction of distillation fractions, (2) the reforming process by fractionating the reformate, (3) naphtha cracking, (4) hydrodealkylation.

1. Benzene (C_6H_6) :

Benzene ranks ahead of toluene, xylene, and naphthalene as seen in the above classification. Except for small amounts used as a solvents and for motor fuels, the major usage of C_6H_6 is for production of phenol and styrene with the remaining approximate 40% divided about equally among other aromatic derivatives such as aniline, sulfonated detergents, DDT, maleic anhydride, chlorobenzene and cyclohexane.

The last chemical, C_6H_{12}, comes from petroleum via two routes : (a) catalytic hydrogenation of benzene, (b) direct fractionation of reforming stock. Cyclohexane is an important intermediate in synthetic fibre preparation. 6, 6-Nylon fibre is produced by oxidation of C_6H_{12} to adipic acid, which is condensed with hexamethy-lenediamine.

2. Toluene ($C_6H_5 \cdot CH_3$) :

Toluene is produced in yields about equal to that of benzene and of xylene in petroleum reforming operations. Unfortunately, the demand for toluene is not as prevalent as for benzene.

The major end uses of toluene are :

- Solvents - paints, rubber and plastic cements.
- Chemicals – benzyl chloride and its peroxide for polymerization catalyst; benzoic acid; as plasticizer; phenol from benzoic acid; toluene di-isocynate in urethane foams.
- Detergents – toluene sulfonates.

These uses cannot absorb all of the toluene production; so plans for using up excess toluene are :

- Blending in motor petrol to upgrade octane number; this produces low income yield as fuel value.
- Production of phenol.
- Dealkylation with H_2 to yield benzene.

The principal source of toluene is catalytic reforming of refinery streams. This source accounts for 87% of the total toluene produced. An additional 9% is separated from pyrolysis of gasoline produced in steam crackers during the manufacture of ethylene and propylene. Other sources are, an additional 2% by-product from styrene manufacture and 1 - 2% entering the market from separation of a coal tar.

3. Xylene [$C_6H_4 \cdot (CH_3)_2$] :

A similar story to that of toluene exists for xylene as each is produced in about equal yields from petroleum refining. Some 60% is used as low-value, high octane blending fuel, the balance of important high-value products are as follows:

- Solvents – high boiling type for alkyd resins.
- Chemicals -
  - (i) From orthoxylene – phthalic anhydride and acid competitive with naphthalene oxidation.
  - (ii) From metaxylene – isophthalic acid which is competitive with phthalic acid for reinforced plastics and plasticizers.
  - (iii) From paraxylene – dimethyl terephthalate used in polyester fibers and films.

4. Naphthalene ($C_{10}H_8$) :

Over 80% of this naphthalene supply to be derived largely from coal will go into the production of phthalic anhydride. The principal market outlets for phthalic derivatives are in alkyd resins and plasticizers with under 10% to dyestuffs and exports. If coal supplied naphthalene is insufficient to meet the demand, hydrodealkylation of petroleum from Indian crude oil can be utilized.

Composition of BTX Aromatics :

Table 5.1 : Composition of BTX

Source of BTX	Composition %		
	B (Benzene)	T (Toluene)	X (Xylene)
Pyrolysis gasoline	56	31	13
Reformate gasoline	10	50	40
Coal tar light oil	77	17	6

Physical Properties of BTX Aromatics :

Table 5.2 : Physical properties of BTX

	Benzene	Toluene	Ethyl benzene	O-Xylene	M-Xylene	P-Xylene
Molecular Weight	78	92	106	106	106	106
Colour & state	Colourless liquid	Colourless liquid	Colourless liquid	Colourless liquid	Colourless liquid	Colourless liquid
M.P. (°C)	5.5	−95.0	−95.0	−25.2	−47.8	13.3
B.P. (°C)	80.0	111.0	137.2	144.4	139.1	138.4
Density (g/cm^3)	0.88	0.87	0.88	0.88	0.87	0.86

5.4 PRODUCTION AND SEPARATION OF BTX COMPOUND [W-13; S-14]

Reformate Separation into BTX Aromatics by Udex Process : [S-16, 17]

Reformate as a feed can be send to the extraction column, where in this column the reformate can be heated to about 140 - 150°C in presence of lean solvent. During extraction we get two phases, one is of extract and another is of raffinate. Extract phase contains the aromatic compounds, while a raffinate phase contains a non-aromatic compound.

Solvent can be used here to extract the aromatics compound from reformate feed and then it can be send to a stripper column. In stripping column, the recovery of solvent take place which can be removed from bottom side, whereas aromatic extract can exit from top side of column. This aromatic extract phase should be cooled and then send to settler. Here two settlers can be used, part of one settler is feed back again to extraction column as a reflux, which is about 1 - 1.5% by volume of feed. By this reflux the percentage recovery increases. Now, the remaining part of aromatic extract phase can be fed to wash tower, for washing with water. Where the dissolved impurities should be removed. Higher percentage of concentrated aromatic extract component can be withdrawn from top as a product, which can contain 99% Benzene and 96% Xylene, where the water with impurity can be obtained from bottom side.

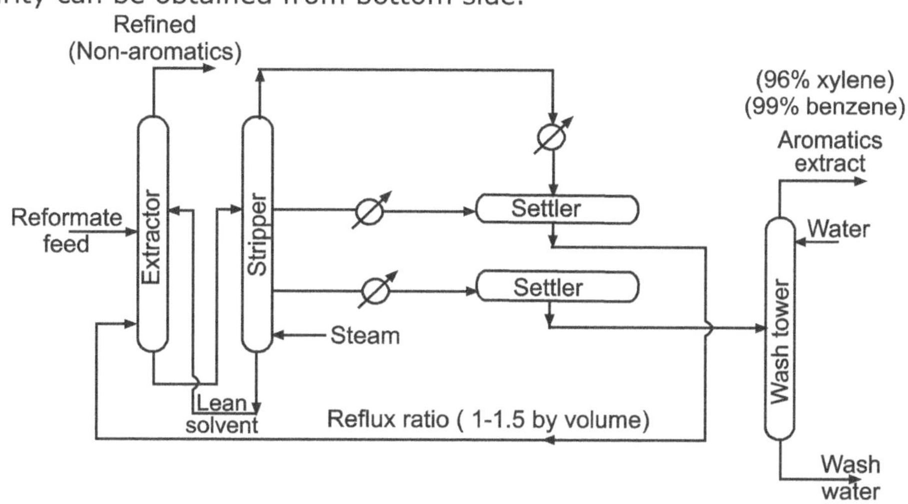

Fig. 5.3 : Udex process for recovery of BTX from reformate gasoline

Separation of BTX into Benzene, Toluene and Xylene :

The aromatic mixture from udex process containing BTX is fractionated in column 1, 2, 3 as shown in Fig. 5.4. The separation is purely guided by relative volatility. This mixture is preheated to about 230 - 270°C and then it can be send to clay tower. Then the bottom effluent can be send to the series of three fractionating column. Where Benzene and Toluene can be obtained as pure components by fractionation. C_8 aromatics consist of ethylbenzene, o, m, p-xylenes and are taken out as overhead product of column 3. The heavier fraction C_9 is obtained as bottom product. The separation of xylenes involves an elaborate procedure.

Further purification of benzene and toluene can be done by fractionation or selective treatments depending upon the purity requirements.

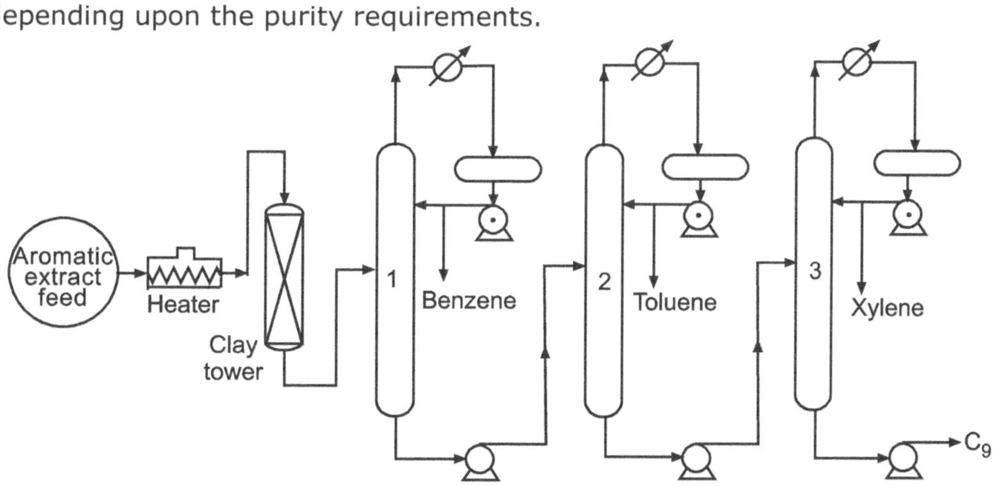

Fig. 5.4 : BTX separation process

5.4.1 Production Obtained from BTX

1. Benzene :

Benzene is a natural constituent of crude oil, and is one of the most basic petrochemicals. Benzene is an aromatic hydrocarbon and the second [n]-annulene ([6]-annulene), a cyclic hydrocarbon with a continuous π-bond. It is sometimes abbreviated as Ph–H. Benzene is a colourless and highly flammable liquid with a sweet smell. Because it is a known *carcinogen*, its use as an additive in gasoline is now limited, but it is an important industrial solvent and precursor to basic industrial chemicals including drugs, plastics, synthetic rubber and dyes.

Today, benzene is used mainly as an intermediate to make other chemicals. Its most widely-produced derivatives include styrene, which is used to make polymers and plastics, phenol for resins and adhesives (via cumene), and cyclohexane, which is used in the manufacture of Nylon. Smaller amounts of benzene are used to make some types of rubbers, lubricants, dyes, detergents, drugs, explosives, napalm and pesticides.

List of products obtained from benzene are ethyl benzene, cumene, cyclohexane, aniline, chlorobenzene, styrene, polystyrene, acetone, phenol, bisphenol, polycarbonates, epoxy resins, phenolic resin, adipic acid, nylon 6-6, caprolactum, nylon 6.

2. Toluene :

It is an aromatic hydrocarbon that is widely used as an industrial feedstock and as a solvent. Like other solvents, toluene is sometimes also used as an inhalant drug for its intoxicating properties.

Toluene is a common solvent, able to dissolve paints, paint thinners, silicone sealants, many chemical reactants, rubber, printing ink, adhesives (glues), lacquers, leather tanners, and disinfectants. It can also be used as a fullerene indicator, and is a raw material for toluene diisocyanate (used in the manufacture of polyurethane foam) and TNT. In addition, it is used as a solvent to create a solution of carbon nanotubes. It is also used as a cement for fine polystyrene kits (by dissolving and then fusing surfaces) as it can be applied very precisely by brush and contains none of the bulk of an adhesives used primarily to make solvents, gasoline additives, and explosives.

List of products obtained from Toluene are Benzene, Toluene diisocynate, Benzoic acid, Polyurethane, Caprolactum, Nylon.

3. Xylene :

Xylene is a chemical that occurs naturally in crude oil but is supplemented by additional amounts made in the refining process. It is used in many common products, such as paints, rubber, adhesives, plastic bottles and clothing.

Xylene encompasses three isomers of dimethylbenzene. The isomers are distinguished by the designations *ortho*- (*o*-), *meta*- (*m*-), and *para*- (*p*-), which specify to which carbon atoms (of the benzene ring) the two methyl groups are attached.

List of products obtained from Toluene are as per their isomer positions :

(a) Meta-xylene : Isophthalic acid, Alkyd resins, Polyamide resins, Unsaturated polyesters.

(b) Para-xylene : Dimethyl terephthalate, Purified terephthalic acid, Polyesters.

(c) Ortho-xylene : Phthalic anhydride.

5.5 MANUFACTURE OF ANILINE FROM PHENOL

(a) Properties of Aniline ($C_6H_5NH_2$) :

Molecular Weight : 93.13

Melting Point : −6.3°C

Boiling Point : 184.1°C

(b) Methods of Production :

- Ammination of Phenol.

(c) Chemical Reaction :

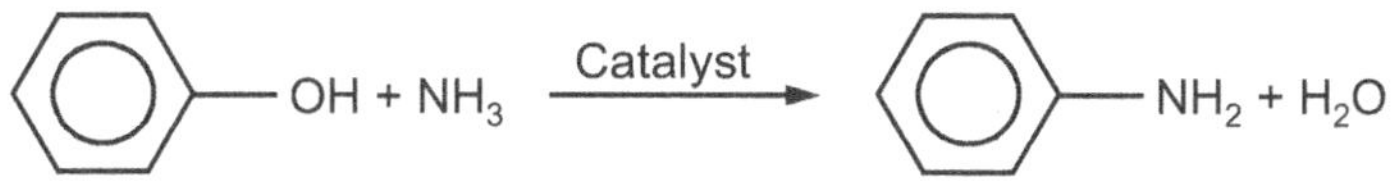

(d) Temperature : 370°C

(e) Catalyst : Silica-Alumina

(f) Process Description :

Animation of phenol takes place in the vapour phase using excess ammonia in the presence of a silica-alumina catalyst at about 370°C and 1.7 MPa. A fixed-bed reactor is suitable, as the reaction exotherm is moderate (about 2 kcal per mol). Use of excess ammonia (mole ratio of 20:1) pushes the reversible reaction to the right and also inhibits the formation of by-products diphenylamine, triphenylamine and carbazole.

The scientific design process for aniline from phenol overcame problems of low yield, slow reaction rate, and poor product recovery that thwarted previous attempts at phenol conversion. One of the key features of the process is the unique SD-developed catalyst which permits almost complete phenol conversion (per pass) at nearly stoichiometric yield of aniline on phenol. As a result, the purification train is very simple and the aniline product is of unusually high purity.

The catalyst has an extremely long life (in excess of seven years). Naturally, this process is favoured by the availability of low-cost phenol feedstock. Aniline is also used as an intermediate for dyes and pigments, explosives, agricultural chemicals and pharmaceuticals.

5.6 MANUFACTURE OF BENZOIC ACID FROM TOLUENE

(a) Properties of Benzoic Acid (C_6H_5COOH) :

Molecular Weight : 122.12

Melting Point : 122.3°C

Boiling Point : 250°C

(b) Methods of Production :

- Oxidation of Toluene.

(c) Chemical Reaction :

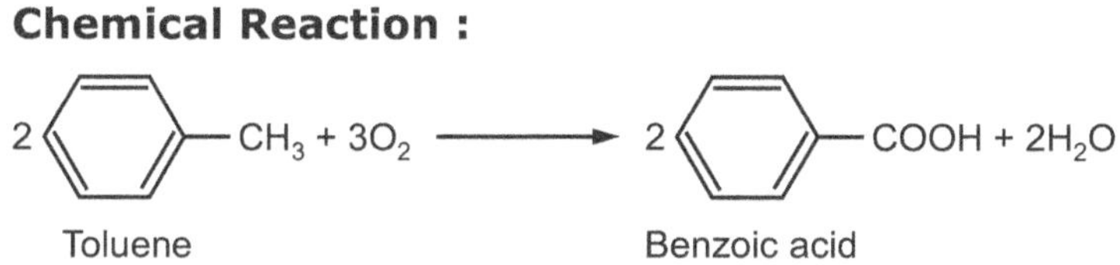

(d) Temperature : 250°F-450°F

(e) Catalyst : Heavy metal oxidation catalyst

(f) Flow Sheet :

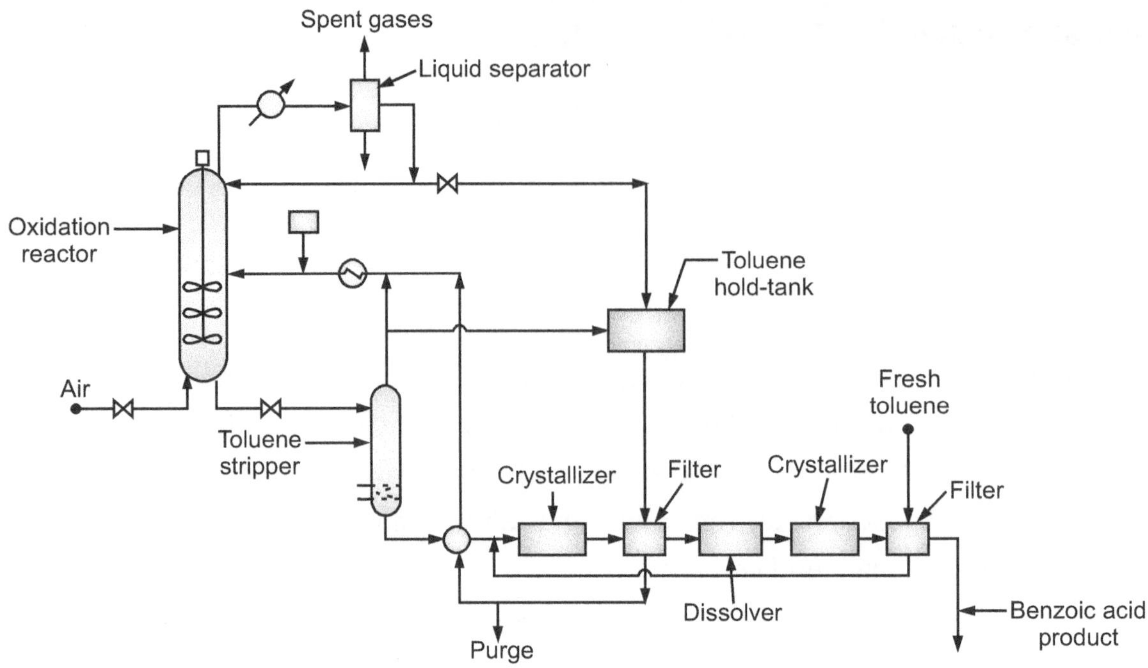

Fig. 5.5

(g) Process Description :

Benzoic acid produced by the oxidation of toluene with an oxygen-containing gas in the presence of a heavy metal oxidation catalyst, for example cobalt or manganese salts, is normally contaminated With non-acidic oxygenated compounds which may be derived from impurities in the feedstock or may constitute the intermediate oxidation products or products of secondary reactions. Such non-acidic materials consist of tars and oxygenated materials such as benzaldehyde, benzyl alcohol, esters of benzoic acid and the like. Contaminants of this type in benzoic acid are highly undesirable since they cause the product generally to be unsuitable with respect to colour, and when present in significant quantities, reduce the value of benzoic acid when employed for subsequent chemical conversion. Among such uses are the preparation of phenol, hydrogenation to hexahydrobenzoic acid, an intermediate for preparation of caprolactam, and preparation of various esters of value as plasticizers, solvent etc.

The object of this invention is to provide an improved commercial oxidation system which is particularly suitable for the conversion of toluene to benzoic acid in such a form that, it may be readily recovered as a substantially pure product. A particular object is to provide a continuous oxidation system which will result in the attainment of product quality and yields comparable or even superior to those attainable in batch processes. A further object is to provide an integrated system for crystallizing and separating benzoic acid from a recycle stream comprising benzoic acid, toluene, catalyst and associated impurities most effectively. These and other objects will be apparent as the detailed description of the invention proceeds.

In practicing the invention, fresh toluene charging stock, together with recycle toluene containing dissolved benzoic acid as hereinafter more fully described, is introduced into the upper part of a continuous oxidation system into which is also introduced oxidation catalyst. The oxygen-containing gas is introduced into the countercurrent system at the opposite end from that at which charging stock is introduced. In some cases, it may be desirable to effect the countercurrent oxidation in a single vessel provided with a plurality of intermediate mixing stages. The heat of reaction is removed by boiling unreacted toluene and water (formed during the oxidation) from the vessel, condensing withdrawn vapours and returning the hydrocarbon condensate after separation of the water. The oxygen concentration in the withdrawn gasiform-stream should not exceed about to 8% in order to avoid possible explosion hazards; the countercurrent system is especially advantageous, in that, it enables practically quantitative oxygen utilization with minimum hazard and without substantial detriment to product yield or quality.

The oxidation of the toluene feedstock is effected in ZIAi, the presence of a heavy metal oxidation catalyst and advantageously in the presence of a catalyst consisting of a heavy metal oxidation catalyst and a bromine affording substance as a promoter. The catalyst metal may advantageously be selected from the group consisting of manganese, cobalt, nickel, chromium, vanadium, molybdenum, tungsten, tin, ceriurn or mixtures thereof, and the metal may be employed in elemental, combined or ionic form, preferably as a compound soluble in the reaction medium. The preferred mixed metal catalysts are mixtures of manganese in the form of bromide or acetate with ammonium molybdate, ammonium chromate, tungstic acid, cobalt acetate.

The oxidation reaction is effected continuously at an elevated temperature, preferably about 350-425°F., and elevated pressure preferably about 200 psi. It has been found that most desirable results, from the standpoint of achieving desired product quality, are obtained when the reactor is operated so as to obtain as an effluent stream a solution of about 20 weight per cent benzoic acid in toluene, although somewhat broader ranges, eg., 15-25% are also effective. The conversion of toluene to benzoic acid in an amount sufficient to yield a reactor effluent of such concentrations is readily accomplished by adjusting the rate of air (or other oxygen-containing gas) introduced to the reactor, as well as the residence time in the reactor. It will be understood that oxygen is fed to the reactor at a rate which will permit rapid consumption of the gaseous oxygen, so as to avoid excessive oxygen content of the off-gases.

The oxidation product from the oxidation reaction is next introduced into a concentrator, which may be a flash chamber or other apparatus suitable for removal of toluene as an overhead stream. The flash chamber or stripper is operated at a temperature of about 300-325°F and pressure between about 1 to 1.68 atmospheric is employed to concentrate the reactor effluent to form about 40-60% by weight, approximately 50% benzoic acid.

The benzoic acid enriched stream from the concentrator is next introduced to a heat exchanger, wherein the benzoic acid containing stream is indirectly contacted with the recycle mother liquor stream obtained from the subsequent filtration step. This heat exchange step offers added economies in process operation, as well as providing a recycle stream for introduction to the oxidation reactor which is at or near reaction temperature. It is essential that, the recycle stream be hot in order to avoid quenching of the continuous oxidation reaction, which is characterized by a certain minimum initiation temperature.

The cooled benzoic acid stream is subsequently adjusted to a temperature between about 110°F and about 130°F, preferably about 120°F, so as to precipitate a substantial quantity of the benzoic acid dissolved therein, and the precipitated acid is separated by centrifugation, filtration or other liquid-solid separating means. All or a portion of the toluene recovered from the concentrat-or is employed as a wash liquor for washing the filter cake, and is recycled with mother liquor to the heat exchanger for subsequent recycle to the oxidation reactor.

When desired, the solid benzoic acid cake so obtained can be further purified by dissolving in fresh toluene and recrystallization there from at a temperature of about 130 - 140°F, preferably about 135°F. Mother liquor from the second crystallization is introduced to the crystallizer for recovery and recycle of dissolved benzoic acid. Finally, the purified benzoic acid product is dried or alternatively melted, stripped of residual solvent, and send it for bagging.

Important Points

- **C_4 hydrocarbon compounds :** Butadiene ($CH_2{=}CH{\cdot}CH{=}CH_2$), MTBE ($C_5H_{12}O$), Butyl Acetate ($C_6H_{12}O_2$).
- **BTX Aromatics :** BTX means Benzene (C_6H_6), Toluene ($C_6H_5{\cdot}CH_3$) and Xylene [$C_6H_4{\cdot}(CH_3)_2$].

Practice Questions

1. Draw the flow sheet for manufacture of ethylene oxide.
2. Describe with neat flow sheet, the manufacturing process of phenol from Isopropanol.

MSBTE Questions and Answers (A Per G-Scheme)

Summer 2016

1. Describe process of butadine manufacture. Draw a flow diagram for the same. **(4 M)**

Ans. Refer Section 5.1.1.

2. Draw a flow diagram for the manufacture of MTBE. **(4 M)**

Ans. Refer Section 5.1.2.

3. Explain Udex process for recovery of BTX from reformat gasoline. **(4 M)**

Ans. Refer Section 5.3.

4. List out two chemicals derived from C_4 hydrocarbons with their uses. **(2 M)**

Ans. Refer Section 5.1.

Summer 2017

1. Draw well labelled flow diagram of manufacture of MTBE. **(4 M)**

Ans. Refer Fig. 5.1.2.

2. Describe UDEX process for BTX recovery. **(4 M)**

Ans. Refer Section 5.3.

3. Write chemical reaction involved in the following manufacturing: **(4 M)**

(i) Butadiene

Ans. Refer Section 5.1.1.

(ii) Butyl acetate

Ans. Refer Section 5.1.3.

4. Name any eight petrochemicals manufactured from benzene, toluene and xylene as a starting material. **(4 M)**

Ans. Refer Section 5.2.1.

5. Explain uses of MTBM. **(2 M)**

Ans. Refer Section 5.1.2.

6. Give two uses of Butadiene. **(1 M)**

Ans. Refer Section 5.1.1.

Winter 2017

1. What is BTX? What are their uses? **(4 M)**

Ans. Refer Section 5.2.1.

2. List the products obtained from C_4 hydrocarbons. State two uses of each product. **(4 M)**

Ans. Refer Section 5.1.

3. Draw the flowsheet for the manufacture of MTBE. **(4 M)**

Ans. Refer Section 5.1.2.

4. Explain the udex process for the recovery of BTX. **(4 M)**

Ans. Refer Section 5.2.

Summer 2018

1. List any four chemicals derived from C_4 hydrocarbon. **(4 M)**

Ans. Refer Section 5.1.

2. Explain the manufacturing of MTBE. **(4 M)**

Ans. Refer Section 5.1.2.

3. What is BTX ? What are their uses ? **(4 M)**

Ans. Refer Section 5.2.1.

4. Explain the manufacture of butadiene with neat flow diagram. **(8 M)**

Ans. Refer Section 5.1.1.